HOM
CHEESEMAKING
2nd Edition

The secrets of how to make your own cheese and other dairy products

NEIL AND CAROLE WILLMAN

Written and Published by Neil and Carole Willman,
Little River, Victoria, Australia.

Printed in Australia by.

National Library of Australia
Cataloguing-in-publication data:

Willman, Neil
Home Cheesemaking

2nd Edition
Includes Index
ISBN   0 646 13342 X
Cheese. 2. Cheese – Varieties. 3. Dairy products. I. Willman, Carole.
II Title.

637.7

Cover photo courtesy of the Australian Dairy Corporation.

The opinions, advice and information contained in this publication have not been provided at the request of any person, but are offered solely for information purposes.

While the information contained in this publication has been formulated in good faith, the contents do not take into account all the factors which need to be considered before putting that information into practice. Accordingly, no person should rely on anything contained herewith as a substitute for specific advice.

CONTENTS

Contents

INTRODUCTION

There are many people who are now making their own cheeses at home. Some of these people have one or more cows or goats for their source of milk, whilst others get milk from a friendly neighbour, or even buy milk from the local shop or supermarket. Cheese can be made from milk from all of these sources, but using bought homogenised milk is only suitable for a limited range of cheeses. Cheese was once made in small quantities in many small factories in Australia, but the bulk of cheese is now produced in large dairy plants. Farmhouse cheesemaking is still very common in European countries and some of these people have migrated to Australia and have started making cheese here. A few of these people have obtained a factory licence and now make and sell their cheese commercially. Most however, make cheese at home for their own consumption.

It is important to note that in Australia you cannot sell your cheese legally unless it is made in a fully licensed and registered premise.

Cheesemaking can bring immense enjoyment and satisfaction to those willing to try. There is a very wide range of cheeses that you can make at home. Some are very simple to make whilst others are not so easy. Under the heading of each recipe you will find a difficulty rating of one, two, three or four stars. One star means the cheese is easy to make, through to four stars being quite difficult and only recommended for the experienced cheesemaker, or those wanting to be challenged. We suggest that you first try an easy cheese such as Quarg and work up from there. Don't think that everything will be plain sailing all the time. You will make mistakes, but you should know that some wonderful new cheeses have emerged from mistakes.

There are a few simple rules to follow when making your cheese. Firstly, all equipment and the hands of the cheesemaker should be meticulously cleaned and sanitised before use. This will ensure that the cheese will have a better flavour and will have a longer shelf life. Next, you should always follow the directions carefully and make a record of the times and temperatures that you use throughout the process. Note anything unusual such as differences in the smell or appearance of the milk, curds or cheese. A good accurate thermometer is an essential tool for the cheesemaker, as temperature control is very important for success.

Most of the equipment needed can be found in the home or can be purchased at a local store.

The basis of cheesemaking is to preserve the milk by three possible methods; firstly, fermentation or souring by the use of lactic acid bacteria, secondly by dehydration or water removal from the milk in the form of whey, and finally by the addition of salt to the cheese.

If you follow the directions carefully you will soon be able to astound your friends when you proudly state, "I made this myself!"

INGREDIENTS

The ingredients needed to make your cheese depend on the type of cheese to be made. Some only require two ingredients, others more. The following is a list of ingredients which may be needed at some time;

- Milk
- Starter
- Adjunct cultures
- Rennet
- Lipase
- Cheese colour
- White Mould spores
- Blue mould spores
- Plastic cheese coating
- Wax
- Herbs and spices
- Salt
- Lemon juice
- Calcium chloride
- Vinegar

WHAT'S IN MILK?

Milk is a truly complex liquid. It is a delicate product that needs careful consideration during handling, storing and cheesemaking. Many people believe that if their milk is liquid and white then it's OK for using to make cheese. This is not the case, as changes take place over time which may make the milk unsuitable for cheesemaking. These changes are not always detectable to our senses.

It is difficult to give an accurate composition of milk because it is never constant. There are so many things that affect its composition of milk. Some of these are outside of your control whilst others are not. The top four items listed at the top of the list play a greater role in affecting the composition than those below them.

Factors affecting milk composition:

- Stage of lactation of the animal (time after last calving)
- Breed of animal
- Feed available to the animal
- Individuality of the animal
- Intervals between milkings
- Age of the animal
- Health of the animal
- Weather conditions

The following table shows the typical composition of cow's, goat's and sheep's milk and the ranges that are possible.

Composition of milk from cow, goat and sheep.

Typical values and the normal range shown in brackets.

Animal	Fat	Protein	Lactose	Salts	Total solids
Cow	4.2 (3.5-6.5)	3.5 (3.1-4.0)	4.9 (4.7-5.1)	0.8 (0.7-0.9)	12.9 (12-14)
Goat	3.6 (3.0-4.0)	3.2 (2.8-3.5)	4.7 (4.0-5.0)	0.8 (0.7-0.9)	11.9 (11.0-13)
Sheep	7.0 (5.5-12.5)	5.7 (4.9-6.8)	4.6 (4.2-5.2)	0.9 (0.8-1.4)	18.3 (16-25)

The protein in milk is comprised of 80% casein and 20% whey proteins. The casein remains in the cheese and the whey proteins are lost in the whey. Ricotta cheese which is made from mainly whey, is an exception and contains mostly whey proteins.

HANDLING MILK

Cooling.

After the milk has been drawn from the animal, it should be refrigerated immediately after milking if it is not going to be used on the same day. It should be stored at 4 to 6°C until used and should not be left for any longer than three days before use. It is possible to make products with milk stored for up to five days old or even longer, but do not expect it to behave the same as fresh milk and do not expect the same quality product. It will more than likely have a defect of some sort. There are bacteria that grow in the milk during refrigeration producing substances known as enzymes. These enzymes slowly break down the protein and cause the curd to form slower and softer. The damaged proteins will probably result in stale flavours.

MILK QUALITY ISSUES

Antibiotics

Under no circumstances should you use any milk from an animal that is being treated with antibiotics. Consult the label on the antibiotics to find the withholding period after treatment has finished. If you do happen to get milk containing antibiotics, your starters will not work and the product will be wasted.

Agitation

Excessive agitation or pumping of milk results in two negatives for cheesemaking. Firstly it incorporates air into the milk which in turn inhibits the action of starters. Secondly it breaks up the fat which could cause rancid flavours in the milk and cheese as well as free fat floating on the surface.

Colostrum

For the first three to five days the secretion is known as colostrum. It should be fed to the newborn animal and never used for cheese.

Mastitis

Cows suffering from mastitis will produce milk with poor cheesemaking properties. The curd formation from rennet will take longer and be softer. The curd will retain whey during the cheesemaking process making it more difficult to produce a hard cheese. The cow suffering from mastitis may appear normal and show no signs of ill health. Such cases are known as subclinical mastitis. The milk from these cows will not produce top quality cheese.

Sediment and extraneous matter

If you are milking your own animal, it would be wise to filter your milk. A colander lined with a fine cloth is ideal as a filter, but you must make sure the milk is still warm. Filtration is best done immediately after milking. A good filter should remove any extraneous matter such as hair, mud, chaff and dust.

Sanitisers and detergents

Care should be taken to rinse all traces of detergents and sanitisers from milking machinery, utensils and containers prior to milking. Residues can cause allergic reactions, taints and inhibition of the starters.

Colour of milk

The colour of milk is usually considered to be off white. Cow's milk is creamier in colour because the milk fat contains carotene giving it a golden colour. Goat's and sheep's milk fat lacks carotene, and hence are whiter. The resultant cheeses will also reflect this by being much whiter than their cow's milk counterparts. After giving birth, the secretion from the animals may be richer in colour.

If the milk is a pink colour it is an indication of blood in the milk, which is normal for the first three to five days after giving birth, but after that time indicates health problems with the animal. Blue coloured milk is a result of the dye used in the intra mammary antibiotics applied to lactating animals with mastitis. Neither milks should be used for cheesemaking or human consumption.

Using homogenised milk

Homogenised milk best suited to fresh lactic cheeses that are curdled by the action of acid and do not need rennet. It can be used for cheeses using rennet provided that calcium chloride is added to the milk prior to rennetting. The results will never be as good as using non homogenised milk, and if the homogenisation process has used very high pressures then it may not be possible to get good curd formation. One millilitre of calcium solution per three litre of milk gives the best results. Homogenised milk even with calcium addition will produce a curd that is more inclined to shatter.

STANDARDISING MILK

What is standardisation?

Standardisation refers to the adjustment of the fat level of the milk. Milk straight from the cow can vary so much in composition, whereas packaged milk is almost always standardised to a fixed composition. The aim of standardisation is to bring the milk to a constant or standard composition. Normally only the fat level in the milk is adjusted.

Why standardise your milk?

It may be necessary to standardise if you wish to make a cheese that is true to its traditional type. For example, to make a hard cheese such as Parmesan, you will need to reduce the fat level in the milk, and to make a creamy Havarti, you will need to increase the fat content. Unless you have access to measuring equipment, you will never be able to accurately standardise the milk, however you can improvise and practise a crude form of standardisation, and accept that you will have some variation in your cheeses.

How can you standardise your milk?

To change the level of fat in the milk you can use a cream separator. A separator is a mechanical device which is very efficient and will remove all of the fat in the form of cream. They are not cheap and therefore suited to large quantities. Once the cream has been separated then you must decide on how much to put back in with the skimmed milk.

Without a separator you can improvise standardisation by allowing the milk to stand overnight or longer in a container in your refrigerator, then skim off some cream from the top of the milk.   This will lower the fat level and give you some fresh cream to use or turn into sour cream, butter or Mascarpone. If you want to increase the fat test to make a creamier cheese, firstly allow the milk to sit in a container that has a tap on the bottom (eg an insulated water container). After settling overnight, some low fat milk can be removed via the tap.

PASTEURISING MILK

Why do we need to pasteurise milk for cheesemaking?

Raw milk may contain undesirable or harmful bacteria. The harmful bacteria are known as pathogens and other undesirable bacteria are known as spoilage organisms.

When making cheese, temperatures used are conducive to the growth of bacteria. If there are any undesirable bacteria in the cheese milk, their numbers will increase during the cheesemaking process. There will be an increased risk of spoiling the final product, or worse, adversely affecting the health of those eating the cheese. It is therefore advisable to always pasteurise milk before making cheese.

Pasteurisation kills all known bacteria that are harmful to man. There is a trend in all developed countries towards the pasteurisation of milk for cheesemaking. The manufacturers need to consider legal liabilities of producing raw milk cheeses and the potential for food poisoning. Pasteurisation also ensures a more uniform bacterial level before commencement of cheesemaking, therefore helping to produce a more consistent cheese. If you intend to make cheese from raw milk, the results will be variable, and may range from magnificent to terrible disasters.

What is pasteurisation?

Pasteurisation is the heat treatment of milk in order to kill off harmful bacteria. There are three accepted methods for pasteurising.

1. Heating to 61°C and holding for 30 minutes will pasteurise your milk.
2. Commercially, milk is heated to 72°C and held for 15 seconds to achieve pasteurisation.
3. Heating to 68°C and holding for one minute will also pasteurise your milk.

For making cheese at home the latter of these options will save you significant time.

How do we pasteurise milk?

You will need to use two containers, the first one to put the milk in, and a second that will be large enough to accommodate the first. A stainless steel bucket or saucepan would be best for the first container. Part fill the larger container with hot water, place on a hot plate and bring to a temperature above 80°C. When the water is hot, place a cake rack in the bottom of the pot, then place the first container with the milk to be pasteurised into the pot of hot water and onto the rack. Stir the milk constantly until the temperature of the milk reaches 68°C, then hold it for one minute before taking the bucket out of the hot water. Place it into a sink of cold water, maintaining the stirring until the temperature reaches that specified in the recipe. An accurate thermometer is essential when pasteurising milk.

The purpose of the rack in the bottom of the larger pot is to ensure that the container of milk does not sit on the bottom of the pot. If it is allowed to contact the bottom of the pot then it will get too hot and cause heat damage to the milk. The milk proteins will be changed in such a way, that after adding rennet the milk may set very slowly or not at all. If it does set, it will most likely be a soft set which is unsatisfactory for making cheese. This is not such a problem if you are making cheeses that don't use rennet, but it is good practice to pasteurise properly in all cases.

An alternative for small quantities is to pasteurise milk in a microwave oven, although with this method you must be careful that you stir the milk frequently, to ensure that the milk on the outside does not boil. This method is good for pasteurising milk for yoghurt and cultured milk, as they require milk heated to 90°C.

Helpful hints to speed up the pasteurisation step.

1. Pasteurise the milk directly after milking while it is still warm.

2. Prewarm cold milk in a sink of hot water before placing into the large pot of hot water.

3. Use a large pot to ensure that the hot water quantity exceeds the milk quantity.

4. Stir continuously during heating

Pasteurising milk in the kitchen

BEHAVIOUR OF GOATS AND SHEEPS MILK IN CHEESEMAKING

One of the main differences between the milks is the fat globule size. The fat exists in milk as small spheres known as globules. The fat globules from sheep and goat's milk are on average smaller and more uniform in size than cow's milk. Cows milk with its larger globules tends to form a cream layer on the top fairly quickly. With goats and sheep's milk the smaller fat globules rise to the top at a slower rate and thus stay more homogenous during storage. Cream can be separated from milk with the use of a mechanical separator, but for the same reasons it is more difficult to extract the cream from goats and sheep's milk.

We have often been asked the question "can any type of cheese be made using goats or sheep's milk? The answer is usually yes. There are some differences in the milk that changes the behaviour during cheesemaking. The milk will usually take longer to coagulate if it is lower in solids. This will apply to all milks, but with sheep's milk being significantly higher in solids it will always set faster than both cows or sheep's milk. Thus for sheep milk the quantity of coagulant (rennet) can be reduced by approximately one third to achieve the same setting time. Goat's milk is typically lower in its solids than cows milk and thus may take slightly longer to coagulate. This is not always the case as we have experienced some goat's milk, which behaved the same as cow's milk.

The two most important factors that affect the solids are the breed of the animal and the quality of the feed. A poorly fed animal will respond by producing low solids milk and vice versa.

During cheesemaking rennet coagulated goats milk tends to lose whey faster than cows milk. The reverse is true for sheep milk, which tends to retain the whey.

The quality of the milk is also an important issue with respect to cheese and yoghurt manufacture. The key steps to produce good quality milk are as follows:

- Quality starts with the health of the animal. All milking animals must be kept healthy and in good condition if good quality milk is to be obtained.

- The next important issue is the cleanliness and hygiene at the milking stage. This means thorough cleaning and sanitation of the milking equipment after each milking and replacement milking equipment rubberware on a yearly basis. The rubbers will become cracked and harbour bacteria which will be then seeded into the milk.

- Milkers hands should also be clean prior to milking and cleaned between animals to reduce any cross infection.

- Milk must be cooled immediately after milking to 4°C.

- Sheep and goats milk are both more likely to contain dirt and sediment from the environment. In any case it is wise to filter the milk straight after milking.

- For cheesemaking, goats milk should be used within 2 days of milking. Delaying longer will result in the breakdown of the fat globules, releasing fatty acids, which will produce a strong goaty flavour.

FREEZING MILK

When you have excessive quantities of milk then freezing is a realistic option. There are some simple rules that must be followed if the milk is to be used for production of cheese or yoghurt at a later stage.

The first golden rule is to freeze the milk while it is still fresh and not to store it for any longer than is necessary. Next it is far better to freeze it in multiple lots of small quantities rather than one large batch. The reason for this is that the freezing must be done quickly. By freezing quickly the ice crystals that form will be smaller and will cause less damage to the protein and fat globules. Therefore to freeze quickly the packs should not be stacked too close together. It stands to reason that a shallow layer will freeze faster than a deep container, thus if you have a choice of containers select the one that will provide the thinnest layer of milk.

The next rule is keeping it cold. The temperature is best at – 18°C, or colder if possible. If stored at higher temperatures e.g. –10°C the protein will become denatured and this will cause some difficulties during further processing.

You may notice a change in colour of the frozen milk. This is normal and it will return to its normal colour after thawing. Thawing can be done slowly or quickly but it should be completely thawed out before removing any milk from the containers. Before removing the milk invert the container a couple of times to mix.

STARTERS

What is a starter?

A starter is not rennet or junket but specially selected bacteria. Starters are chosen because they possess certain characteristics, which are of value to cheesemaking. These characteristics can include such things as acid production, enzyme production, flavour component production or gas production (carbon dioxide). Some starters have one of these characteristics, while others have more than one. Starters are very tiny. In one millilitre of liquid starter there are about 200,000,000 (two hundred million) living starter organisms.

Why do we use starters?

Starters are mainly used for acid production. Many years ago naturally occurring bacteria in the milk were used, however their numbers are inconsistent and many of the bacteria in raw milk are spoilage organisms. The quality of cheeses made from raw milk will be quite variable. Pasteurisation kills many of the desirable bacteria in raw milk and hence the need to add starters to replace them.

Acid production is necessary in the making of cheese, as the acid helps the curd to form and shrink, as well as preventing the growth of spoilage bacteria and determining the characteristics of the cheese.

Starter culture characteristics

Throughout the recipes in this book, the starter types have been given alphabetical codes. This is to avoid the use of lengthy bacterial names and stop any confusion that may arise.

Type A Starter

This is a medium temperature (mesophilic) starter, consisting of *Lactococcus lactis subspecies cremoris* and *Lactococcus lactis subspecies lactis*. (Now you may understand why we have chosen alphabetical codes)! It is selected when acid production only is required. If you are preparing your own starters, they should be incubated at a temperature between 22 and 30°C until the milk curdles (~12-20 hours). The optimum is about 30°C, but it can be grown at a lower temperature to delay the setting of the culture. They may be used for the production of Cheddar, Fetta, Camembert, Blue Vein, Cottage Cheese and Quarg.

Type B Starter

This is a medium temperature (mesophilic) starter flavour type, consisting of *Lactococcus lactis subspecies cremoris, Lactococcus lactis subspecies lactis, Lactococcus lactis subspecies lactis biovar diacetylactis* and with or without one or more *Leuconostoc* species. If you are preparing your own starters they should be incubated at a temperature between 22 and 30°C until the milk curdles (~12-20 hours). They are selected because of their acid, flavour and gas production. They are used for the production of Edam, Gouda, Camembert, Havarti, Tilsit and other cheeses. They are also used for the production of sour cream and cultured butter.

Type C Starter

There are two categories of type C starter, those for cheese and those for yoghurt. The strains for yoghurt are selected because of their ability to produce polysaccharides, which enhance the yoghurt viscosity. The strains for cheese do not have this characteristic. Type C starter is a high temperature (thermophilic) starter, consisting of *Streptococcus thermophilus* and either *Lactobacillus delbreukii subspecies bulgaricus* or *Lactobacillus helveticus*. If you are preparing your own starters they should be incubated at a temperature of 37°C until the milk curdles (~6-8 hours). They are selected for their acid and flavour production. They are used for making yoghurt and many Italian cheese varieties.

Type C aBt Starter

This is a cocktail of the *Lactobacillus acidophilus*, **B**ifidobacterium species, *Streptococcus thermophilus and maybe Lactobacillus bulgaricus*. It is a popular blend to make yoghurt that has a significant population of the acidophilus and Bifido cultures. To obtain maximum levels of the probiotic cultures the yoghurt milk is incubated at 38°C until the milk curdles (~6-8 hours).

Type D Starter

Type D starter consists solely of strains of *Lactobacillus acidophilus*. It is used for making Acidophilus milk and is best grown at 38°C until the milk curdles (~18-24 hours).

Type E Starter

Type E starter consists of specially selected strains of *Streptococcus thermophilus*, which are acid sensitive. It is thermophilic and when incubated should be kept at 37°C until the milk curdles (~6-8 hours). The culture is used in the modern version of Camembert and Tikmilk, a delightful cultured milk drink.

Type F Starter

This is a culture for making a special type of cultured milk similar to yoghurt, but containing *Lactobacillus acidophilus* and *Bifidobacterium* species. The product made from these cultures is very popular in Scandinavian countries and has a number of health claims made about it. These are naturally occurring bacteria found in the human digestive tract. They are incubated at 38°C until the milk curdles (~18-24 hours). These should be used directly into the milk. Due to their long incubation times, they cannot be effectively propagated as starters. If other bacteria get into the milk they may outgrow the starter, causing spoilage.

Starter in Direct Vat Set (DVS) on the left and right, prepared liquid starter (centre).

HOW TO PREPARE STARTERS FOR CHEESEMAKING

Starter may be obtained in freeze-dried or Direct Vat Set form. If you are using Direct Vat Set (DVS) cultures, you do not need to prepare liquid starter. DVS cultures are in powdered form and can be added directly to milk when making cheese. However it is very difficult to accurately measure out the exact quantity of DVS powder when using small quantities of milk. Quantities of powder supplied are usually for hundreds of litres of milk. It is therefore difficult to use exactly the right quantity if you have less than 50 litres of milk. By using the DVS powder to make up liquid starter, the correct amount of liquid starter can be easily measured in a measuring jug. You can make up liquid starter as shown on the chart on page 31, using either freeze dried or DVS starter. Making liquid starter is also a more economical way of using starter.

Starter Types CaBt, D and F for Acidophilus and Acidophilus/Bifido cultured products must be used directly into the milk as directed in the recipe and should not be subcultured. If you choose to prepare your own liquid starter follow the instructions below. Only starter types A, B, C and E should be subcultured this way.

Note that any jars or utensils used for the preparation of starters MUST be sterilised before use. Either place jars into boiling water for a few minutes, leave them in the oven while cooking or sanitise them with a chemical sanitiser such as sodium hypochlorite. See sanitising notes for more details.

1. Prepare enough cool boiled milk for your starter needs. This can be done by bringing some milk to the boil on the stove or in a microwave oven, then allowing it to cool to room temperature whilst covered. An alternative is to use UHT milk. Place a small amount of starter powder into some cool boiled milk. The amount of starter powder is not critical and only affects the time the milk takes to coagulate.

2. Put the lid on the container. Swirl the milk to mix in the powder.

3. Incubate the milk by keeping it at the appropriate temperature, i.e. 20 to 25°C for type A and B starters, and 37°C for type C and E. If you use the higher end of the temperature range then the milk will set quicker and vice versa. You can use this fact to suit your own timetable.

4. Once the milk sets, it should be kept in the fridge until used. You should use it within 24 hours.

5. To enable you to make up more starter, as soon as the milk sets you can remove a small amount with a sterilised spoon, and place it into some more cool boiled milk. If you wish, this can be stored in the fridge for up to one week, and brought out for incubation the day before your next cheesemaking.

6. Add coagulated starter to the cheese milk at the appropriate stage in the recipe.

Is the starter DVS i.e. suitable for adding directly into the milk for cheesemaking?

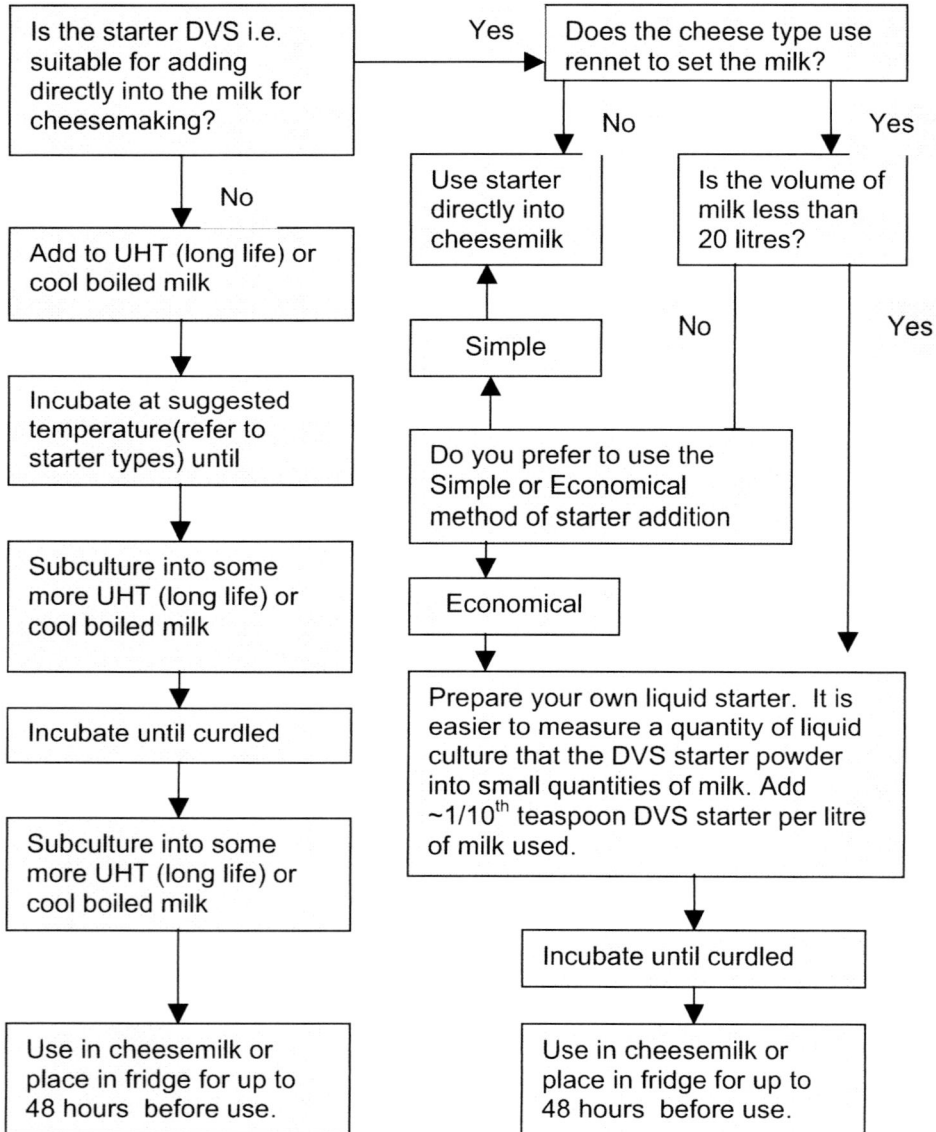

— Yes → Does the cheese type use rennet to set the milk?

No ↓

Add to UHT (long life) or cool boiled milk

↓

Incubate at suggested temperature(refer to starter types) until

↓

Subculture into some more UHT (long life) or cool boiled milk

↓

Incubate until curdled

↓

Subculture into some more UHT (long life) or cool boiled milk

↓

Use in cheesemilk or place in fridge for up to 48 hours before use.

No ↓ (from "Does the cheese type use rennet to set the milk?")

Use starter directly into cheesemilk

↑

Simple

↑

Do you prefer to use the Simple or Economical method of starter addition

↓

Economical

↓

Prepare your own liquid starter. It is easier to measure a quantity of liquid culture that the DVS starter powder into small quantities of milk. Add ~1/10th teaspoon DVS starter per litre of milk used.

Yes ↓ (from "Does the cheese type use rennet to set the milk?")

Is the volume of milk less than 20 litres?

No → / Yes ↓

↓

Incubate until curdled

↓

Use in cheesemilk or place in fridge for up to 48 hours before use.

Step by step guide to preparing starter

After having made your liquid starter you can subculture it a number of times as shown in the following chart. Subculturing is the process of propagating more starter, by using the prepared starter you have made as your beginning point for the next batch.   Subculturing is not suitable for starter types B, C, D, and F.

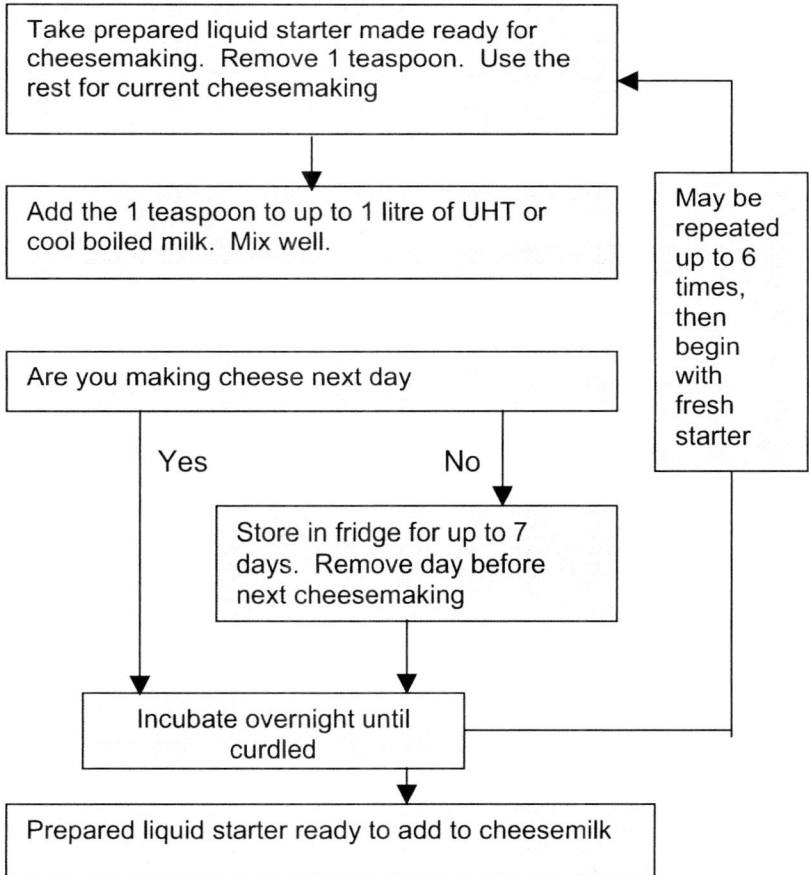

```
┌─────────────────────────────────────────┐
│ Take prepared liquid starter made ready  │
│ for cheesemaking.  Remove 1 teaspoon.    │◄──────┐
│ Use the rest for current cheesemaking    │       │
└─────────────────────────────────────────┘       │
                    │                      ┌────────────┐
                    ▼                      │ May be     │
┌─────────────────────────────────────┐   │ repeated   │
│ Add the 1 teaspoon to up to 1 litre  │   │ up to 6    │
│ of UHT or cool boiled milk. Mix well. │   │ times,     │
└─────────────────────────────────────┘   │ then       │
                    │                      │ begin      │
                    ▼                      │ with       │
┌─────────────────────────────────────┐   │ fresh      │
│ Are you making cheese next day        │   │ starter    │
└─────────────────────────────────────┘   └────────────┘
       Yes              No
        │                │
        │                ▼
        │   ┌──────────────────────────┐
        │   │ Store in fridge for up to 7│
        │   │ days.  Remove day before  │
        │   │ next cheesemaking          │
        │   └──────────────────────────┘
        │                │
        ▼                ▼
┌──────────────────────────────┐
│ Incubate overnight until       │
│ curdled                        │
└──────────────────────────────┘
                │
                ▼
┌──────────────────────────────────────────────────┐
│ Prepared liquid starter ready to add to cheesemilk │
└──────────────────────────────────────────────────┘
```

Maximising the use of your starters

OTHER CULTURES

Mould Spores

Mould spores are sometimes used in the manufacture of cheese. These are available in two categories; Blue mould spores (*Penicillium roqueforti*) and White mould spores (*Penicillium candidum*). Blue mould spores are used for making Blue Vein, Roquefort, Stilton and Gorgonzola cheeses. White mould spores are used for making Camembert and Brie.

Brevibacterium linens, Yeast, White mould powder and Blue mould suspension.

Gas producing starter

These include a bacteria (*Propionibacterium shermanii*) added to the milk to provide extra gas production to form the eyes in Swiss cheese.

Aroma cultures (adjuncts)

A bacterial culture *Brevibacterium linens* is used for rubbing on the surface of washed rind cheeses such as Tilsit and Havarti. *Geotrichum candidum* and yeasts are also used to create different effects on the colour, flavour and aroma.

Lactobacillus casei is used for both cheesemaking and as a probiotic (beneficial) bacterium in cultured milk products. This bacteria has the ability to produce acid but its production is very slow making it necessary to add it with other more active starters.

RENNET

Rennet is an enzyme, which reacts with the protein in the milk to coagulate the milk. It is therefore added to the milk to turn it from the liquid state to a solid state. The active enzyme is known as Chymosin. Rennet is not a starter. Calf rennet (extracted from the calf stomach) is widely seen as the best form of rennet for cheesemaking.

When using rennet always dilute it in 10 times its volume of cool boiled water. If you don't do this you will find that your rennet has started working before you have mixed it in properly, giving a very uneven curd. Be sure that the water you are using is not hot or it will adversely affect the activity of you rennet.

Do not sanitise rennet measuring utensils with sodium hypochlorite. Sodium hypochlorite will adversely affect the activity of your rennet. You will need to rinse these utensils with boiling water.

Hard cheeses cannot be made without rennet. Hard cheeses claiming to be rennet free are more correctly 'animal rennet' free. These cheeses are made from microbially produced chymosin, which is classed as fermented or vegetarian rennet.

As a guide, it takes two and a half millilitres of rennet to set 10 litres of milk in 30 to 40 minutes, if the milk is at 31°C. If the temperature is lower, then it will take longer or if it is too cold it may not set for hours and will be unsatisfactory. If the temperature is 37°C the milk will set in 20 to 25 minutes. Never add more than two and a half millilitres to 10 litres of milk if making a cheese that is going to be stored for any longer than a month. The reason for this is that excess quantities of rennet can lead to a bitter flavour in the cheese. It should not be necessary to add more than three millilitres of rennet per 10 litres of milk for making any cheese.

Rennet should be stored under cool, dark conditions to maximise its strength and life. A normal refrigerator is ideal, and the rennet should last for at least one year with only five percent loss in activity. It will slowly weaken during storage, so it may be necessary to slightly increase the quantity added to the milk, to achieve the same setting time, but do not increase the level above three millilitres of rennet per 10 litres.

Avoid using over heated milk with rennet. Rennet will not coagulate UHT and overpasteurised milk.

Rennet in an easy to use liquid form.

The rennet on the left is single strength calf rennet (145 International Milk Clotting Units per millilitre). On the right is quadruple strength microbial rennet (690 IMCU/mL).

CUTTING THE CURD

In most cheese recipes you will be required to perform the step of cutting the curd. It is not possible to accurately describe the state of the curd when it is ready to cut. The degree of firmness can vary between cheeses but most have a level much the same. The curd will form slowly and firm up as time passes. It should not be cut at the first sign of coagulation but at a later stage when it has firmed up enough so as to be able to cut relatively easily without breaking up on cutting. A good test to use is as follows; take a sharp knife or a thin spatula and make a vertical cut into the curd about six centimetres deep and about ten centimetres long. Remove the knife, turn it on its side, and insert it into the cut under the length of the cut then lift upwards gently. The cut should open into a V shape and some whey will start to run into the V. This whey should be a clear greenish colour, not milky and the curd should not collapse on the knife. If in doubt leave it a little longer. If left for too long, it will be too firm and will be difficult to handle during the rest of the process.

Once the curd is ready to cut, insert a long knife into the container as shown in the diagram, making two sets of vertical cuts at right angles to each other, before cutting in on an angle to make a horizontal cut. The curd should always be left to sit for five to ten minutes after cutting to allow the curds to firm up prior to any stirring. On commencement of stirring, any oversized particles can then be cut with the knife. You can expect to find some larger pieces that come up from the bottom.

Some square or rectangular cake racks may be suitable for curd cutting some soft cheese types. Such racks should have the wires running in one direction only. They are most suited when cutting curd to 1.5 –2 cm cubes.

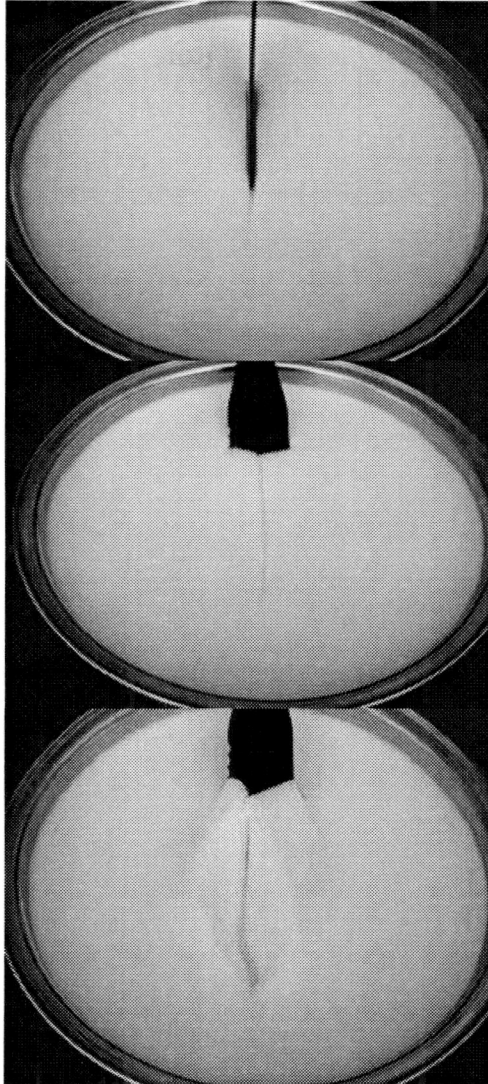

Sequence of testing the curd to assess readiness for cutting

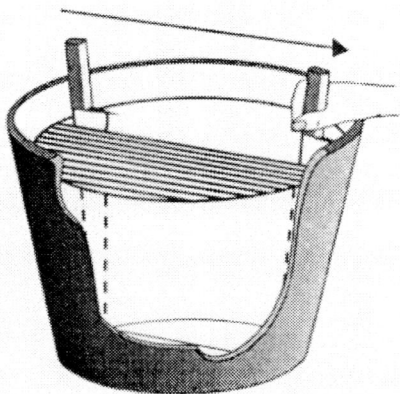

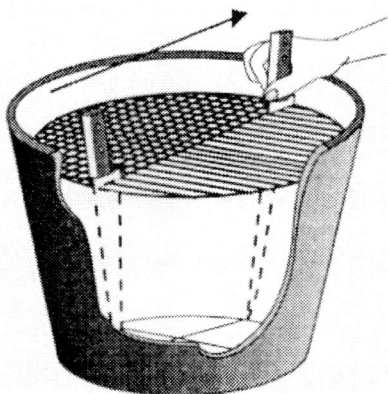

Cutting the curd requires two vertical cuts at right angles to each other, and then a final horizontal cut.

STIRRING THE CURDS AND WHEY

Agitation or stirring of the curds and whey is often required when making cheese. The stirring serves to keep the curd particles apart and helps some whey to come out of the curd. As a rule hard cheeses requires quite a lot of stirring during this stage, and softer cheeses need less stirring.

Stirring, if required by the recipe, must be done gently at the beginning and can be increased in intensity as the curds become firmer. Do not be too vigorous when stirring or the curd will break apart. If your whey becomes milky it may indicate too vigorous stirring.

Stirring can be achieved by using any large flat kitchen utensil such as an egg turner or a perforated skimmer.

COOKING THE CURDS AND WHEY

The process of heating the curds and whey is commonly known as cooking, even though the heating may only raise the temperature a few degrees. For example, in cheddar cheesemaking the cooking involves raising the curds and whey from 32 to 38°C. In other cheeses it may only mean an increase in temperature of one or two degrees.

In some recipes cooking is not required, but normally the setting temperature must be maintained at the specific temperature as stated in the recipe. Keeping the correct temperature is essential if the starter is to do its job.

The temperature can be adjusted or maintained by either direct or indirect means. Each recipe will include a cooking step if it is necessary. Unless otherwise stated, cook the curd by one of the indirect methods.

Direct heating

In recipes for cheeses such as Gouda, Edam, Havarti and some others the curds and whey are heated by direct addition of warm or hot water. Take particular note of the quantities to be added. The water should not be added all at once but over a 15 - 20 minute period. The water will not only heat the curds but the amount of water will also help influence the flavour of the cheese.

The reason for this is that some of the lactose inside of the curd is washed out. The starter turns lactose into lactic acid, and hence a curd with less lactose will have less lactic acid and a milder flavour.

If you are making one of the above cheeses and wish to modify the flavour, then you could adjust the amount of water added. If you add more water then the cheese will have a milder flavour and vice versa. If more water is added then its temperature should be lower to achieve the same overall temperature. Likewise if less water is added its temperature should be higher.

Indirect heating

Here are three ways that you can either cook the curds or maintain the temperature.

- The cheesemaking vessel can be placed in a sink of warm water, with constant replacement of some of the water with hot water.

- The cheesemaking vessel can be placed in a preserving unit and use the unit's own thermostat.
- You can use some basic kitchen equipment such as a large baking dish and frying pan. Water is put into the frying pan and a large baking dish or a large stockpot is placed on top of the frying pan. The pot then has a cake cooling rack placed in the bottom and the cheesemaking vessel is placed onto the rack.

It is best to have the water surrounding the cheesemaking vessel about one or two degrees above the milk setting temperature. This can be maintained or increased when needed by occasionally turning up the thermostat on the frying pan to boil the water.

SALTING

Salting of cheese is an important step in its manufacture, and despite the temptation to make your cheese without salt, it is not advisable. It may be possible to reduce the salt content, but there could be adverse consequences in the cheese. Salt helps to control the growth of undesirable bacteria and hence preserves the cheese. It also helps to promote some changes in the cheese during storage as well as enhancing the flavour.

There are two main methods of incorporating salt into the cheese, dry and brine salting. Dry salting can be achieved by either rubbing salt into the surface of the cheese or by sprinkling salt over the curds before hooping. Brine salting is carried out by placing the whole cheese into a salt/water solution.

As a rule most cheeses that are brine salted are placed into brine which contains between 20 and 26% salt. Brine is saturated at 26.4% salt, and if more salt is added, it will not dissolve and will settle on the bottom. To make up brine at 20% strength, take 200 grams of salt and make up to one kilogram with boiling water. Allow to cool. It is best to acidify the brine slightly by adding five millilitres of vinegar to each litre of brine.

It is not easy to work out how long to brine salt a cheese, as there are many things that affect the speed that salt diffuses into the cheese. The most significant of these is the size and shape of the cheese. If you have two cheeses both weighing two kilograms, one in the shape of a flat disc e.g. Brie and the other in a wheel shape, eg. baby Gouda then the Brie will take up the salt in half the time of the baby Gouda.

Other factors that affect the brine salting include:

- Moisture content of the cheese. A softer higher moisture cheese absorbs salt quicker than firmer drier cheese.
- Brine temperature. Warmer brine means faster salting.
- Cheese temperature. Warmer cheeses means quicker salting.
- Surface condition of the cheese. If the surfaces of the cheese are completely closed then salt will diffuse at a steady rate. If there are openings from the surface into the body of the cheese then the salting rate may be up to 10 times faster and the cheese may become too salty if left in the brine for the normal time.
- Time of immersion. Doubling the time of brine salting does not mean doubling the salt level in the cheese. As

time passes the rate of salt absorption into the cheese slows down.

Brine should not be discarded after use. It can be kept for years if treated correctly. To maintain brine properly does not take a lot of effort. It should be kept to a strength of not less than 20% and filtered of any extraneous matter including particles of cheese regularly. If the brine is well maintained it should only need to be boiled once a year.

Problems with brine salting

Slimy cheese surface:- If the salt content of the brine is too low, bacteria will grow in the brine causing the surface of the cheese to become slimy. It may also turn slimy if the pH of the brine is not adjusted before use. This can be done by adding a little vinegar to your brine.

Cracking cheese rind:- Can be caused by brining in a warm saturated brine.

WAXING CHEESE

Hard cheeses that are going to be stored for more than one month will dry out and crack if not protected. They may also become mouldy and inedible. These cheeses may be waxed before maturing to prevent them from drying out. The waxing will be more successful if you apply a plastic cheese coating to the cheese first. This gives the wax a fat free and moisture free surface on which to adhere.

Two coats of plastic cheese coating are normally applied by brushing with a small paint brush, or rubbing the coating on with a coarse sponge. Coat half of the cheese first, then allow to dry before completing the second half of the first coat. Then repeat for the second coat. Allow the second coat to dry before waxing. To wax the cheese, the wax must be heated up until it is all molten. Holding one end dip the cheese into the wax for five to ten seconds. Remove and allow to dry. This should only take a couple of minutes. Once the wax is dry, turn the cheese over and wax the other end. The wax temperature will determine the thickness of the coat. If the coat is too thin then allow the wax to cool a little, and then dip again. The optimum is between 90 and 120°C.

Half of the cheese is immersed into the wax, removed for two and a half minutes to dry then repeated for the second half.

ASH COATING

The use of ash to coat cheese is a typically European technique. The ash may be medicinal carbon or ash produced in a charcoal burner. The technique involves the cheeses being made into its final form followed by a light coating with ash over the entire cheese surface. The effect of the ash is to soften the flavour a little by absorbing some of the acid and cheese flavours. A French cheese known as Morbier has an ash layer horizontally applied through the centre of the cheese.

OTHER COATINGS

The use of other coating to the surface of the cheese is one way to bring about a different effect either visually or by altering the flavour.

Coatings such as grape seeds, vine leaves, hop leaves, food colourings, spices, wine washing are some suggestions.

MATURING (STORAGE) CHEESE

The storage of cheese during the maturing stage of cheesemaking is probably the most difficult part of the process. The storage requirements for cheese can be broken down into a few possibilities.

1. Fresh lactic curd such as Quarg, Cottage, Ricotta and Mascarpone need to go directly to refrigerated storage. The ideal temperature for storage is 4°C. The shelf life is short i.e. from a few days to a couple of weeks.

2. Cheeses such as Cheddar, Gouda, Edam, and Havarti require storage between 8 and 12°C. The best storage options depends on the ambient conditions. A cellar or a second refrigerator set at maximum may do the trick. Some fridges will hold 11°C on the highest setting.

3. Some fridges can be split by placing an insulating layer eg polystyrene/cardboard above, and thus warming the lower vegie crisper section for cheese storage.

4. Storage of white mould ripened cheeses requires not only storage at cool temperatures (see point 2 above) but also humidity levels of ~95%. To maintain humidity, the cheese can be placed on a non-corrosive rack in a storage container with the base of the container covered with water and lid on tight.

5. To obtain a humid cool climate an entire esky could have its base covered with water and racks placed inside to support the cheese. Ice bricks are then used to control the temperature.

6. Cheese such as Parmesan once dried can be stored at higher temperatures without problems. Temperatures around 15°C are ideal.

7. A cellar may be ideal for maturing cheeses. Foreign moulds and insects may spoil the cheese if left unprotected.

CHEESE YIELD

(How much cheese you should get from your milk)

The biggest disappointment to home cheesemakers is to see the extent of the curd shrinkage as the whey is removed from the curd. The expected yield from a few common cheese types is shown below:

Cheese Type	Yield
Cheddar	10%
Gouda	10%
Romano	8.5%
Parmesan	7.5%
Camembert	12-14%
Fetta	13%
Quarg	20%

It is the milk which holds the key to the maximum possible yield of cheese. Milk rich in solids will yield more cheese. If are you standardising the milk by removing some cream, for every 100 mL of cream removed you will lose about 80 grams of cheese.

Next comes the cheese moisture content. If the moisture content of your cheese is lower then this means you have taken more whey out of the cheese and of course the yield will be adjusted down accordingly. Leaving more whey in the curds will give more yield but may mean that the cheese is too moist, thereby reducing its shelf life.

EQUIPMENT NEEDED

The following is a list of equipment, which may be required at various times, depending on the cheese variety being made.

- *Cheese vat*

For small quantities of milk you can use any food grade plastic tubs or buckets, as long as they can be heated. Do not use any plastic containers with cracks, as the cracks will harbour bacteria. We have used plastic ice-cream containers, for quantities as small as two litres. Stainless steel pots or pans and buckets are ideal. In most cases, when the curds are to be kept warm or heated it is a good idea to use two pots, where one is large enough to fit inside the other.

- *Measuring jars and spoons*

An assortment of different size of jars and spoons always comes in handy.

- *Cheesecloth*

Cheesecloth is needed to drain bag cheeses such as Quarg and Cream cheese, and also for placing on boards or trays when used for turning or draining cheese in hoops.

- *Curd knife*

For small quantities of up to 10 litres a long sharp kitchen knife is fine. For larger quantities, you may need to get stainless steel frames made which are strung like a harp, one with horizontal nylon lines and the other vertical.

- *Ladles*

Flat perforated ladles come in handy for removing Ricotta from the pot, transferring curds hygienically and also for stirring.

- *Moulds or hoops*

Hoops are the containers that the curds are placed into to provide the final shape of the cheese. Galvanised tins or cans may be used provided they don't rust. It is quite common for PVC storm water pipe to be cut to length and used as a hoop. Diameters up to 150 mm are commonly available at hardware or plumbing stores. An extensive range of specially made plastic hoops is also available for use.

A selection of hoops and baskets for moulding cheese

• *Thermometer*

A thermometer that will range from 0° to 100°C is essential. We prefer to use an easy to read dial thermometer with stainless steel stem. Graduations of 1°C are essential.

• *Colander*

A colander is useful for separating curds and whey, draining Ricotta as well as other uses.

• *Bowls*

Various sized bowls will always come in handy during cheesemaking

• *Syringes*

A small syringe will enable you to measure out small quantities of rennet accurately. Sterile syringes can be purchased from most chemists.

• *Timer*

A timer is very useful for reminding you when to carry out the next step. Not adhering closely to times can change the nature of the cheese.

• *Stirring equipment*

Various kitchen utensils may be used.

• *Wax bath*

A wax bath is useful if you intend to make hard cheeses. The wax can be placed into an old saucepan or a tin that can withstand heating on a hot plate.

• *Cheese press*

A homemade press can be put together quite simply. The
following diagram shows a simple wooden cheese press.

A simple cheese press.
Weights are placed on top to press the cheese.

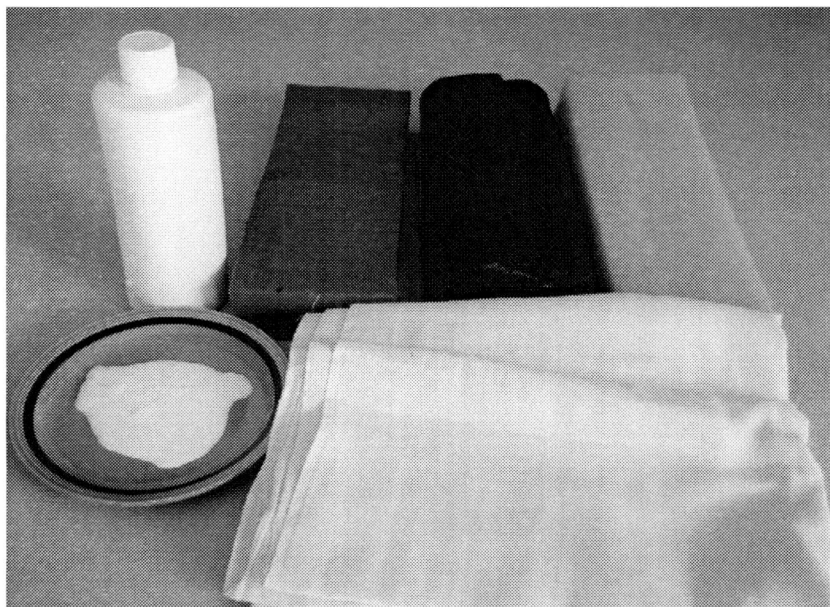

Plastic cheese coating, cheese wax and cheesecloth

- *Acidity measuring equipment*

  If you are serious about cheesemaking and wish to improve your quality control over the process, then you should invest in the basic titration equipment and standard solutions available from laboratory supply companies. Another handy piece of equipment is a pH meter.

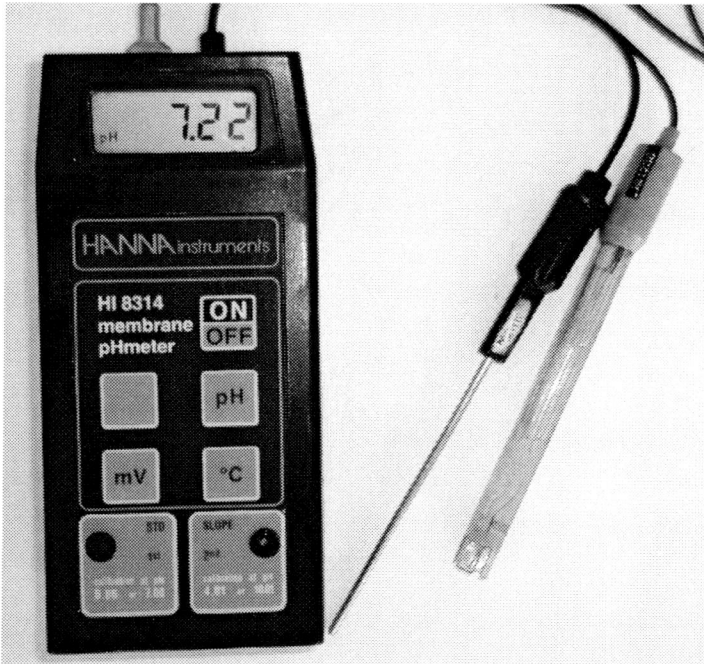

A battery operated pH meter, temperature probe and pH electrode

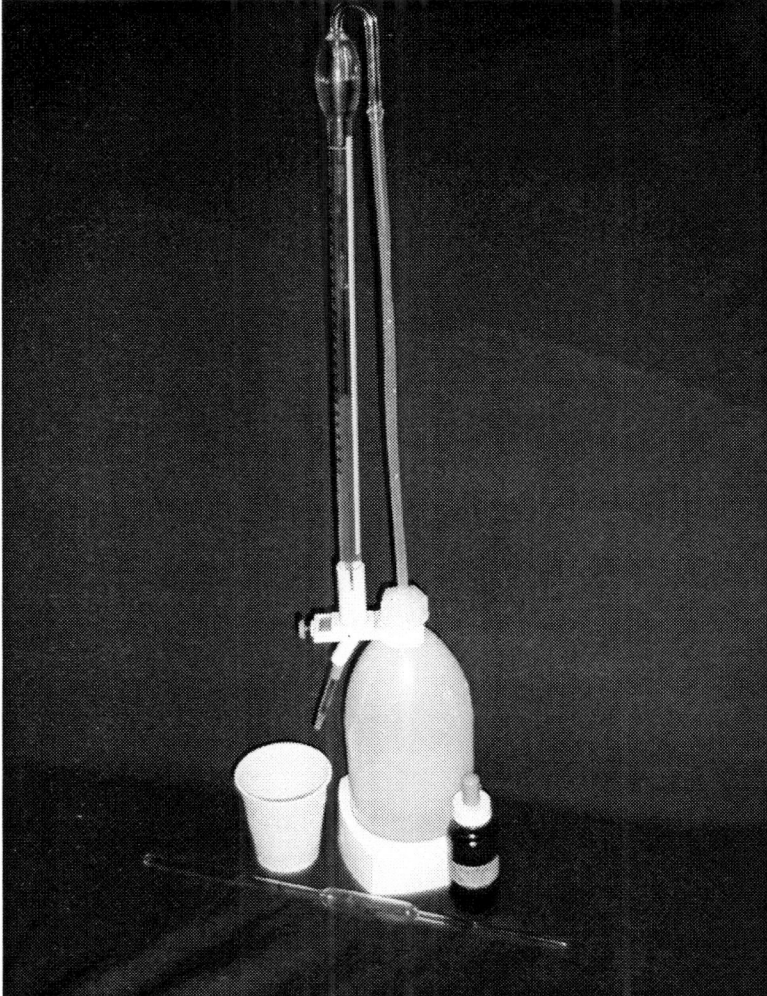

Acidity testing equipment: Test cup, pipette, burette and indicator (in eye dropper bottle)

ACIDITY and pH TESTING

If you get serious about your cheesemaking and want to achieve consistency of your cheese for commercial or other reasons then you would be well advised to start testing either the pH or acidity. The two tests are completely different but both are valuable to the cheesemaker. Some cheesemakers prefer to measure both whilst others chose one or the other.

The acidity test.

The acidity test measures the titratable acidity (TA) of liquids only. The final result is expressed as a percentage or parts per hundred. The acidity of fresh milk ranges from 0.14% to 0.20%. Values close to 0.16% are most common. Sheep milk values are likely to be 0.20% or slightly higher. The acidity test is a useful test as a crude indicator of milk quality. If the milk is typically 0.15% TA for example and the next batch you use is 0.18% then it is likely that bacteria have been growing and producing acid resulting in the increase in titratable acidity. The test is simple but care must be taken to obtain accurate results.

Equipment and materials required

- Clean white cup, beaker or container about 100 – 200 mL.

- Burette graduated to 0.1 mL but preferably 0.05mL.

- 0.1Normal Sodium Hydroxide (NaOH) solution. (added to the burette)

- 5% strength phenolphthalein indicator.

- Pipette capable of delivering 9 mL of sample.

Acidity test procedure.

The sample should be about 20 – 25°C but can be tested at any temperature with only a very small error in the result.

1.   Pipette 9 mL of the sample into the white container.

2.   Add 4 drops of 5% phenolphthalein indicator solution.

3.   Titrate the 0.1N NaOH solution into the container swirling constantly to mix. Continue adding the NaOH until a faint pink colour appears.

4.   Note the reading on the burette. Record the mLs used.

5.   Divide the number of mLs by 10 to obtain the acidity reading.

Example: If the volume used was 1.55mLs then the titratable acidity of the sample is 1.55/10 = 0.155%

The pH test

The pH test measures the concentration of hydrogen ions in the solution. The hydrogen ions are what makes acids acidic. Yes, it's a bit technical but it's a useful test because it's possible to measure the pH of both liquids and solid curds.

The pH test requires the use of some scientific equipment and great care must be taken to protect the equipment and to get accurate best results.

Equipment and materials required

- pH meter and electrode

- Standard solutions (known as buffers). Usually a 4 and 7 buffer is satisfactory. These are used to calibrate the electrode and meter.

  - Tissues
  - Rinse water
  - Test cups

pH test procedure.

It is not possible to give specific instructions regarding pH measurement, as there are many different models available, each with its own set of instructions.

Before a test is performed the electrode and meter must first be buffered (calibrated). The procedure is carried out according to the supplier's instructions, but usually involves calibrating the system on the 7 buffer first followed by the 4 buffer. The electrode must be rinsed and dabbed dry with tissue paper between measurements.

When measuring pH there are a few important points to note:

- Clean the electrode after use as recommended.
- Avoid damage to the glass electrode
- Store the electrode in the recommended storage solution when not in use.
- Calibrate the unit at least daily when making cheese.

pH values indicated in the recipes in this book are indicative only and may vary from the stated values

CLEANING AND SANITISING

Detergents are cleaning compounds whereas sanitisers are chemicals that destroy most bacteria in a given time and render the sanitised surfaces safe from harmful bacteria. Sanitisers are not a substitute for good hygiene, they merely complement your good hygienic practices. All equipment that is going to come into contact with the cheese milk must be cleaned then soaked in a sanitising solution prior to use. An exception to this is equipment coming into direct contact with rennet. See the rennet section for more details. The hands of the cheesemaker should be thoroughly cleaned, then dipped in the sanitising solution before and during the cheese making process.

The most common sanitiser used in cheesemaking is Sodium hypochlorite, commonly called 'hypo' in the cheese industry. Household bleaches with their active ingredient as sodium hypochlorite are suitable. Do not use products that have both sodium hypochlorite and sodium hydroxide as their active ingredients. Always use cold water to make up the sanitising solution, as hot water will destroy the activity of the sanitiser.

A fresh solution should be made up every day that you make cheese to ensure sufficient sanitising action. It may be necessary to change the solution during the day, if it becomes soiled or cloudy. Equipment to be dipped into the sanitising solution should be cleaned first before sanitising.

To make up a sanitising solution:

For a bleach containing 4% Sodium hypochlorite, add 6 millilitres of bleach to each litre of cold water.

To sanitise equipment for starter preparation, use 12 millilitres of bleach to each litre of cold water.

Sterilising materials may also include chlorine based tablets used for sterilising babies bottles.

Never use disinfectants containing Quaternary Ammonium Compounds (such as Pine O'Clean) when cheese making. The active constituents in these sanitisers are strongly bacteriocidal, and even the tiniest residue left in any containers can affect the activity of the starters.

CONTAMINANTS

Cheese can be contaminated with Yeasts, Moulds, E coli, Staphylococcus or other spoilage organisms. The effect of these contaminants could range from off flavours in the cheese, to the production of toxins. If you have pasteurised your milk some of these contaminants may be indicators of poor hygiene in the manufacturing process. The contamination can also be due to airborne contaminants, which may be difficult to eliminate.

Bacteriophage

Bacteriophage, otherwise known as phage, is a virus which can kill starter bacteria. It can cause problems if making cheese every day. If making cheese daily, it may be necessary to occasionally change the strain of the cultures used. It is unlikely to be a problem unless you make cheese on a daily basis and have poor cleaning and sanitation.

RECIPES - CHEESE

BLUE VEIN
****

There are a number of Blue Vein cheeses made around the world. Blue Vein in Australia and the United States is a named variety which is a cow's milk version of Roquefort. Roquefort is a French cheese made from raw sheep's milk. All Blue Vein cheeses need to have the mould *Penicillium roqueforti* added to create the unique colour and flavour. The cheeses are pierced with needles before maturing to allow air into the cheese to enable the mould to grow. Copper wires were once used for this task, but were discontinued when stainless steel became available.

1. Pasteurise the milk and bring it to a setting temperature of 33°C. Add 100 mL of prepared Type B starter for each 10 litres of milk used, i.e. 1%. Mix in well and add blue mould.   pH ~ 6.6 - 6.7

2. Add rennet at a rate of two and a half mL for each 10 litres of milk. Dilute the rennet with at least 10 times its volume of cool boiled water, i.e. 25 mL of cool boiled water to each two and a half mL of rennet. Pour the diluted rennet immediately into the milk, taking care to pour it over as much of the surface as possible, stirring all the time while pouring it in. Mix in well for no less than one minute and no more than three minutes. Maintain the setting temperature until step 7.

3. Allow the milk to set.   This should take 35 to 40 minutes.

4. Cut the curd into 12 mm cubes. Allow the curds and whey to sit for five minutes.   pH ~ 6.4 – 6.5

6. Gently stir the curds and whey to turn the curd over once every five minutes to prevent matting, for a total of 60 minutes.

Gently stirring curds for blue vein.

6. Stir for 15 minutes, then drain all of the whey from the curds.

7. Gently stir the curds by hand for five minutes to prevent matting. Sprinkle one tablespoon of salt over the curds and gently mix in.

8. Place an open ended hoop onto a cheesecloth lined draining tray and place the curds into the hoop. Turn over at half an hour, 1 hour, 2, 4 and six hours.

9. Next day rub salt onto the cheese surfaces and leave for 24 hours.   pH ~ 4.6-4.8

10. Turn the cheese over and repeat the salting then leave it for a further 24 hours.

11. Wipe off any excess salt.

12. Cool the cheese to 8°C and punch holes into cheese, about 20 mm apart. Use an ice pick or thin knitting needle. It is best that the needle does not have a sharp point as the holes will close up afterwards. The end should be partly blunted, but not fully blunt.

13. Mature the cheese at 12 to 14°C for two weeks then at 7 to 8°C for three months. The humidity should be kept high during this time. The excess surface mould can be removed after one month and the cheese wrapped in foil until ready to eat. After wrapping in foil the cheese can be placed in a plastic bag and the bag folded under the cheese during storage. This will help maintain high humidity around the cheese.

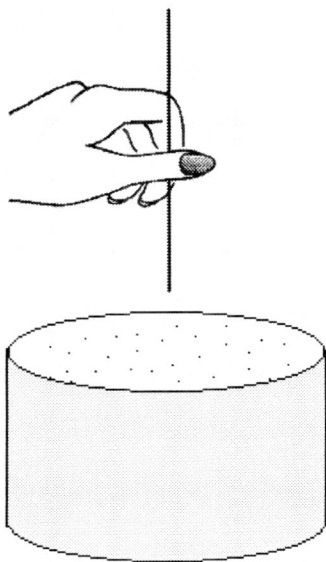

Piercing of blue cheese

BLUE and WHITE MOULD COMBINATION
****

This type of cheese has become quite popular as it combines the characteristic of a white mould cheese which softens the blue cheese flavour intensity. The blue mould is added to the milk and is thus distributed through the entire cheese. The secret to getting the blue and white mould separate is to immerse the newly made cheese in very hot water to kill the blue mould spores on the surface and then follow with a heavy spray of a white mould suspension.

1. Pasteurise the milk and bring it to a setting temperature of 33°C. Add 100 mL of prepared Type B starter for each 10 litres of milk used, i.e. 1%. Mix in well and add blue mould. pH ~ 6.6 –6.7

2. Add rennet at a rate of two and a half mL for each 10 litres of milk. Dilute the rennet with at least 10 times its volume of cool boiled water, i.e. 25 mL of cool boiled water to each two and a half mL of rennet. Pour the diluted rennet immediately into the milk, taking care to pour it over as much of the surface as possible, stirring all the time while pouring it in. Mix in well for no less than one minute and no more than three minutes. Maintain the setting temperature until step 7.

3. Allow the milk to set. This should take 35 to 40 minutes.

4. Cut the curd into 12 mm cubes. Allow the curds and whey to sit for five minutes.

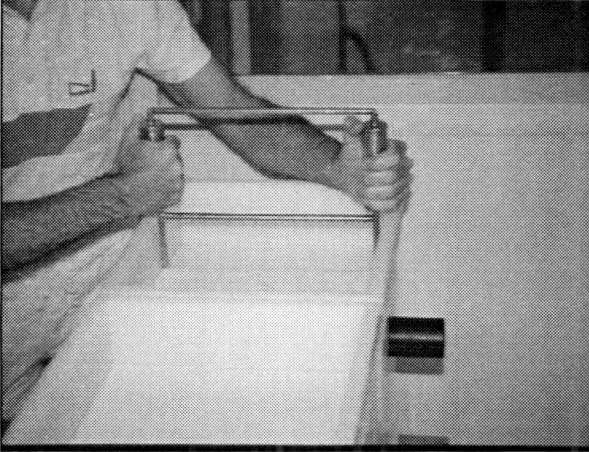

Cutting the curd is easy with specialised cutters!

5. Gently stir the curds and whey to turn the curd over once every five minutes to prevent matting, for a total of 60 minutes.

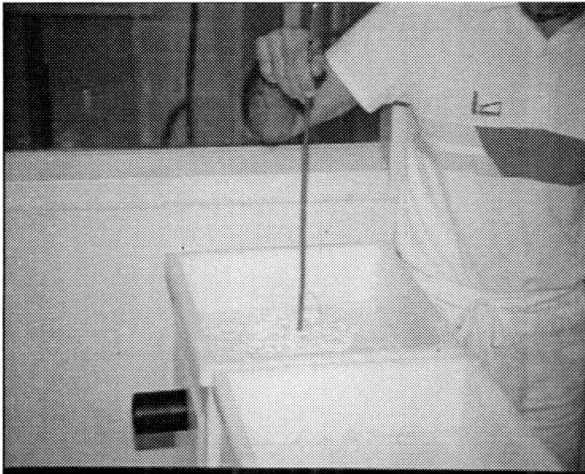

Stirring a 35 litre batch of Blue-White cheese curds using a plunger.

6. Stir for 15 minutes, then drain all of the whey from the curds.

7. Gently stir the curds by hand for five minutes to prevent matting.

8. Place an open ended hoop onto a cheesecloth lined draining tray and place the curds into the hoop. Turn over at half an hour, 1 hour, 2, 4 and six hours.

    pH ~6.0 –6.2

9. Next morning, take the cheese from the hoops and place into a cold 20% brine solution. Here they remain for 20 minutes to one hour, depending on the size of the cheese. To make the brine, add 200 gm salt to 800 mL of boiled water and allow to cool.

    pH ~ 4.8

10. Remove from the brine.

11. Place on a rack to dry overnight

12. Cool the cheese to 12°C.

13. Use an ice-pick or thin knitting needle to spike holes right through the cheese at approximately 1.5 cm intervals.

14. Immerse the newly made cheese in 90-95°C hot water for 5 seconds.

15. Make a suspension of white mould using cool boiled water and white mould.

16. Spray the solution liberally over the cheese. Turn cheese over to ensure all sides are done.

*A simple garden sprayer (sanitised) may be used
providing it has not been used to spray chemicals.*

17. Mature the cheese at 12 to 14°C for two weeks (the
    humidity should be kept high during this time), then wrap
    the cheese in foil (shiny side towards the cheese).
    Maintain the cheese at 7 to 8°C for four weeks prior to
    consumption.

BOCCONCINI
***

Bocconcini cheese is a unique Italian cheese that is used as a fresh eating cheese. It is ideal when eaten with fresh tomato. It obtains its stretchy, stringy texture during a step where the curds are immersed into a bath of hot water, and are worked until smooth. It is a cheese that has a very mild flavour and should be eaten fresh. People that don't like Bocconcini complain that it is tasteless and bland. Whist this may be true a well made Bocconcini has a delicate milky flavour much appreciated by many people.

1. Pasteurise the milk and bring it to a setting temperature of 40°C. Add 150 mL of prepared Type C starter for each 10 litres, i.e. 1.5%. Mix in well.

2. Add rennet at a rate of one and a half mL for each 10 litres of milk, diluting it with at least 10 times its volume of cool boiled water, i.e. 15 mL of cool boiled water to each two and a half mL of rennet. Then pour the diluted rennet immediately into the milk, taking care to pour it over as much of the surface as possible, stirring all the time while pouring it in. Mix in well for no less than one minute and no more than three minutes. Maintain the setting temperature.

3. Allow the milk to set. This should take 50 to 50 minutes.   pH ~ 6.5 – 6.6

4. Cut the curd into 25 mm cubes, then let stand for 30 minutes before stirring.

5. Stir very gently intermittently over the next 60 minutes. The amount of stirring determines the firmness/softness of the cheese. Overstirring will make it too firm.

6. Drain off the whey.   pH ~ 6.1 – 6.2

8.　Keep the curd warm at 40°C. This will cause the curd to fuse together into a single mass. Turn the mass of curd over every 15 minutes to keep it warm. Whey will continue to come out of the curd and should be drained off at regular intervals.

9.　Stretch test. About one hour after removing the whey, take a small thin slice of curd and place into water at 70°C. When the curd is warmed through, remove it from the water and work it well with the fingers. Using the thumb and forefinger of both hands, stretch the curd until it breaks. If the curd is ready to stretch it will be very elastic. If the curd is brittle, it is not ready to stretch. (If you have pH test strips, the curd will stretch when the pH is between 5.0 and 5.4).

10.　If the curd is not ready to stretch, continue to keep the curd warm (as in step 8) and do a stretch test (as in step 9) every 15 minutes until it is ready.

11.　When curd is ready to stretch, cut the curd into very thin slices, and place into hot water at 70°C. You may find it easier to place the sliced curd into a colander, as you can easily raise and lower the curd in and out of the water. To stretch the curd, first work the curd with your hands until there is no openness in the curd and it is quite elastic. You will need to pick up the curd in your hands and pull it apart. At this stage it should not break, but should stretch out. You should then fold it together and repeat the stretching, until all the curd has been well worked. Overworking the curd will also toughen it, as will water that is too hot.

12.　Using such a high water temperature, you will need to wear thick rubber gloves while working the curd or you will burn your hands! You can use two wooden spoons or ladles instead, but we find using our hands gives a better feel of the texture of the curd.

13. Shape the cheese into the desired form by squeezing a small quantity between your thumb and forefinger then squeezing tight to pinch and separate the moulded portion.

14. Place the cheese into ice cold pasteurised water, to set the shape that you have just made. The water may have a little salt added if you prefer but the salt should net be detected in the final cheese.

15. The cheese is ready to eat immediately, or can be stored for up to a week in the fridge.

BRIE
***

To make a Brie the cheese needs to be a large flat disc at least 20 centimetres in diameter. Because of the size and shape of the Brie it ripens faster when nearing maturity. You will need about 7 litres of milk to make a 1 kg Brie.

1.   Pasteurise the milk and bring it to 32°C. Add 200 mL of prepared Type B starter for each 10 litres, i.e. 2%. Add about 1/10th of a teaspoon of white mould spore powder.. Mix both in well. Leave for 60 to 90 minutes. The amount of mould powder is not critical.

2.   Add rennet at a rate of two and a half mL for each 10 litres of milk. Dilute the rennet with at least 10 times its volume of cool boiled water, i.e. 25 mL of cool boiled water to each two and a half mL of rennet. Then pour the diluted rennet immediately into the milk, taking care to pour it over as much of the surface as possible, stirring all the time while pouring it in. Mix in well for no less than one minute and no more than three minutes. Maintain setting temperature until step 7.

3.   Allow the milk to set. This should take 30 to 35 minutes.

4.   Cut the curd into two cm cubes.   pH ~ 6.4 - 6.5

5.   Allow the curd to sit for five minutes.

6.   Turn all the curd over gently once with large flat ladle or similar utensil.

7.   Repeat step 6 twice, at 10 minute intervals.

8.   Drain off one third of the whey, and replace with warm pasteurised water. The mixture should now be at 35°C. (This step is optional.)

9.   Repeat step 6.

10. Drain off half of the whey and pour the remaining curd into 20 cm diameter hoops. The hoops should be placed onto a draining tray lined with cheesecloth.

11. Invert the hoops after 10 minutes and again after half an hour and then at three, five, and eight, hours. This can best be done by having a second cloth lined tray placed on top of the hoops, then firmly holding both trays, turn over.

12. Leave overnight.

13. Next morning, take the cheese from the hoops and place into a cold 20% brine solution. Here they remain for two and a half hours. To ensure even salting turn the cheese over half way through the brining process. To make the brine, add 200 gm salt to 800 mL of boiled water and allow to cool. Add a speck of mould powder to the cooled brine then refrigerate to chill before use. pH ~ 4.6-4.8

14. Remove from the brine and place on a wire rack to dry for 24 hours at room temperature.

15. Put the cheese into a humid environment at 11 to 15°C and store for 8 to 10 days. The cheeses need to be turned on day 3, 6 and 8. The cheese should be fully covered with mould by the end of this time.

16. Place the cheese in foil wrap, or plastic wrap and store for another one to two weeks at 11 to 15°C.

17. The cheese should be ready to eat in three to six weeks after making.

CAERPHILLY
***

Caerphilly is a cheese originating from Wales and has been described in differing ways. One description has it as a cheese of uniform size, neat in appearance and weighing about 3.5 kg, although earlier versions of half a kilogram and two kilogram were known. It is described as a cheese that should be covered with an even coat of white mould, has a white firm body, clean flavour and a short texture. The surface is rubbed with a mixture of rice flour and barley meal to give a uniform finish. Caerphilly originated in the early 1800's in a small town of the same name, just north of Cardiff and means 'castle town'. Another description makes no mention of the white mould and states that the cheese has a pure white body, is moist and smooth with a fresh salty flavour and is eaten at two weeks of age.

To make a standard size cheese of 3.5 kg about 30 to 35 litres of milk is needed.

7.   Pasteurise the milk and bring to a setting temperature of 32°C. Add 200 mL of prepared Type A starter for each 10 litres, i.e. 2%. Mix in well. Leave for a ripening period of 30 minutes.

8.   Add rennet at a rate of two and a half mL for each 10 litres of milk. Dilute the rennet with at least 10 times its volume of cool boiled water, i.e. 25 mL of cool boiled water to each two and a half mL of rennet. Then pour the diluted rennet immediately into the milk, taking care to pour it over as much of the surface as possible, stirring all the time while pouring it in. Mix in well for no less than one minute and no more than three minutes. Maintain the setting temperature.

9.   Allow the milk to set. This should take 30 to 35 minutes.

10.   Cut the curd into 1 cm cubes, then allow the curds and whey to stand for five minutes before stirring.

11.   Five minutes after cutting stir gently for 15 minutes then heat up to 35°C (stirring whilst heating) over the next 15 minutes. Stir for another 15 minutes or until firm enough. This can be determined by breaking a piece of curd in two. Each part should retain its shape.

12.   Allow the curd to settle to let the acidity develop. The time required for this will vary according to starter activity, but 30 minutes should be enough. Run the whey off.

13.   Pile the curd onto the sides of the vat to allow the whey to drain from the curd. The rough outer edges can be trimmed with a knife and the trimmings placed on top of the pile. Allow the curd to drain for approximately 30 minutes, then cut the curd into finger length strips and pile on the side of the vat. The acidity should now be allowed to continue to develop for another 45 minutes before cutting into two and a half cm cubes.

14.   Salt should be sprinkled onto the curds and stirred in. The amount of salt is one gram per litre of milk used.

15.   Place the curd into a cheesecloth lined hoop, leave for 20 minutes, then press for one and a half hours. Remove the cheese from the press and pull the cloth to remove any wrinkles. Return the cheese and press again until the next morning.

16.   Remove the cheese and place into a 20% brine solution. A 3.5 kg cheese should be brined for 24 hours. Smaller cheeses require less time.

17. Dry the cheese and place onto a clean dry board.

18. To prevent the cheese from drying out too much it should be placed into a vacuum sealed plastic bag, or have plastic cheese coat applied, then waxed. Alternatively if a white mould surface is desired on the surface of the cheese, there should be some white mould spore powder placed into the brine before brining the cheese, and the cheese should be stored in a cool damp, high humidity area. The white mould should appear on the surface in about 7 days. The cheese may be stored on either a wire rack or to give a different effect, a board.

19. Store at 11 to 15°C for two weeks.

CAMEMBERT (NORMANDY STYLE)
***

Camembert is one of the most famous French cheeses. In 1791 Marie Harel, a farmer's wife in Camembert perfected a method for making the cheese and developed a marketing scheme for her cheese. Camembert type cheese was known of well before this, but it is formally credited to Madame Harel. Camembert and Brie are both ripened by a luxurious growth of white mould on their surface. They are ready to eat within one month of production.

1. Pasteurise the milk and bring it to 32°C. Add 200 mL of prepared Type A or B starter for each 10 litres, i.e. 2%. Add about $1/10^{th}$ of a teaspoon of white mould spore powder.. Mix both in well. Leave for 60 to 90 minutes. The amount of mould powder is not critical.

2. Add rennet at a rate of two and a half mL for each 10 litres of milk. Dilute the rennet with at least 10 times its volume of cool boiled water, i.e. 25 mL of cool boiled water to each two and a half mL of rennet. Then pour the diluted rennet immediately into the milk, taking care to pour it over as much of the surface as possible, stirring all the time while pouring it in. Mix in well for no less than one minute and no more than three minutes. Maintain setting temperature until step 7.

3. Allow the milk to set. This should take 30 to 35 minutes.

4. Cut the curd into two cm cubes.   pH ~ 6.4 -6.5

5. Allow the curd to sit for five minutes.

6. Turn all the curd over gently once with large flat ladle or similar utensil.

7. Repeat step 6 twice, at 10 minute intervals.

8. Drain off one third of the whey, and replace with warm pasteurised water. The mixture should now be at 35°C.

9. Repeat step 6.

10. Drain off half of the whey and pour the remaining curd into hoops. The hoops should be placed onto a draining tray lined with cheesecloth. pH ~ 6.1 – 6.2

11. Invert the hoops after 10 minutes and again after half an hour and then at three, five, and eight, hours. This can best be done by having a second cloth lined tray placed on top of the hoops, then firmly holding both trays, turn over.

12. Leave overnight.

13. Next morning, take the cheese from the hoops and place into a cold 20% brine solution. Here they remain for 20 minutes to one hour, depending on the size of the cheese. To ensure even salting turn the cheese over half way through the brining process. To make the brine, add 200 gm salt to 800 mL of boiled water and allow to cool. Add a speck of mould powder to the cooled brine then refrigerate to chill before use.
       pH ~ 4.6-4.8

14. Remove from the brine.

15. Place on a wire rack to dry for 24 hours at room temperature.

16. Put the cheese into a humid environment at 11 to 15°C and store for 8 to 10 days. The cheeses need to be turned on the third, sixth and eighth day. The cheese should be fully covered with mould by the end of this time.

17. Place the cheese in foil wrap, or plastic wrap and store for another one to two weeks at 11 to 15°C.

18. The cheese should be ready to eat in three to six weeks after making.

Camembert curd draining in the hoops

CAMEMBERT (MODERN)
**

This version of camembert is far simpler to make and is a good way to begin if you plan to make Camembert. It does not have quite the same character as the Normandy version but is still very pleasant. It can be consumed after only two weeks but improves over the next four weeks.

1.   Pasteurise the milk and bring it to 40 to 42°C. Add 200 mL of prepared Type E starter for each 10 litres, i.e. 2%. Add about $1/10^{th}$ of a teaspoon of white mould spore powder. Mix both in well. Leave for 45 minutes. The amount of mould powder to add is not critical.
$$pH \sim 6.6 - 6.7$$

2.   Add rennet at a rate of two and a half mL for each 10 litres of milk. Dilute the rennet with at least 10 times its volume of cool boiled water, i.e. 25 mL of cool boiled water to each two and a half mL of rennet. Then pour the diluted rennet immediately into the milk, taking care to pour it over as much of the surface as possible, stirring all the time while pouring it in. Mix in well for no less than one minute and no more than three minutes. Maintain setting temperature until step 7.

3.   Allow the milk to set. This should take 30 to 35 minutes.

4.   Cut the curd into two cm cubes . $pH \sim 6.4 - 6.5$

5.   Allow to sit for 30 minutes.

6.   Turn all the curd over gently, for three minutes

7.   Allow to sit for 30 minutes.

8.   Turn all the curd over gently again as in step 6.

9.   Allow to sit for 30 minutes.

10.   Turn all the curd over gently again as in step 6.

11.   Allow to sit for 30 minutes.   pH ~ 6.0 – 6.3

12.   Drain off half of the whey and pour the remaining curd into hoops. The hoops should be placed onto a draining tray lined with cheesecloth.

13.   Invert the hoops after 10 minutes and again after half an hour and then at three, five, and eight, hours. This can best be done by having a second cloth lined tray placed on top of the hoops, then firmly holding both trays, turn over.

14.   Leave overnight.   Next morning pH ~ 5.0 – 5.2

15.   Next morning, take the cheese from the hoops and place into a cold 20% brine solution for 20 minutes to one hour, depending on the size of the cheese. A cheese of 125 grams needs about 20 minutes, and a 250 gram cheese needs about one hour.(To   make   the brine add 200 gm salt to 800 mL of boiled water and allow to cool. Place brine in the refrigerator to chill the brine before use. Add a speck of mould powder to the cooled brine).

16.   Remove from the brine.

17.   Place on a rack to dry for 24 hours at room temperature.

18.   Put the cheese into a humid environment at 11 to 15°C and store for 8 to 10 days. The cheeses need to be turned on the third, sixth and eighth day. The cheese should be fully covered with mould after 10 days.

19.   Place the cheese in foil wrap, or plastic wrap and store for another one to two weeks at 11 to 15°C.

20.   The cheese should be ready to eat in two to four weeks after making.   pH ~ 7.0

Camembert cheese completely covered in white mould

CHEDDAR CHEESE FARM STYLE
***

One of the great things to come from England is Cheddar cheese. It is now made in many other countries throughout the world. It originates from the Cheddar region in the south of England and was perfected and standardised by Joseph Harding in the middle of the nineteenth century. Cheddar cheese is still made in Cheddar Gorge today.

The dairy industries of Australia, New Zealand and the USA have all embraced cheddar as a product of major economic significance. Almost all cheddar is now made in large fully automated production plants. The product is rindless cheddar. Quite a few farmhouse cheese plants in the UK still produce the traditional rinded cheddar, but in Australia there are no more than a handful of traditional Cheddar makers.

Before you rush into making a Cheddar, note that it requires a minimum of three months to obtain a mild flavour and at least a year to get a vintage flavoured cheddar. The longer the cheese has to mature, the more time there is for flavours to develop. Don't expect it to be perfect every time. Even the commercial manufacturers can have problems getting the flavour to exactly what they want. They usually grade the cheese at three weeks and three months of age, to see if it is suitable for further storage or to be sold straight away.

The characteristic body of cheddar comes about from the cheddaring step, where the curds are kept warm for about 120 minutes to allow the unique structure to develop.

1. Pasteurise the milk and bring it to a setting temperature of 32°C. Add 200 mL of prepared Type A starter for each 10 litres, i.e. 2%. Mix in well.

    pH ~ 6.6 – 6.7

2. If a deeper colour is desired for the final cheese annatto colouring may be added. The addition rate ranges from 0.5 mL per 10 litres for a slight lift in colour, up to 5 mL for a deep red colour.

3. Add rennet at a rate of two and a half mL for each 10 litres of milk. Dilute the rennet with at least 10 times its volume of cool boiled water, i.e. 25 mL of cool boiled water to each two and a half mL of rennet. Then pour the diluted rennet immediately into the milk, taking care to pour it over as much of the surface as possible, stirring all the time while pouring it in. Mix in well for no less than one minute and no more than three minutes. Maintain the setting temperature.

4. Allow the milk to set. This should take 30 to 35 minutes.

5. Cut the curd into one cm cubes, then let stand for five minutes before stirring.   pH ~ 6.5 – 6.6

6. Over the next two and a half hours, the curds and whey have to be stirred. Stir gently at first, and then more vigorously as the curd becomes firmer. As you start stirring you should heat the curds and whey up to 38°C. The time for heating should take 40 to 45 minutes.

7. Seventy minutes from the time you cut the curd, drain off whey until half of the original milk volume is reached. This is within the two and a half hours mentioned in step 5.   pH ~ 6.4 – 6.5

8.   Continue stirring intermittently until two and a half hours has elapsed from the time of cutting the curd. All this stirring is to drive the whey out of the curds to make your cheese firm enough. Failure to get the curd at the correct firmness at this stage will result in a soft curd cheese or a hard crumbly cheese.

$$pH \sim 6.1 - 6.2$$

9.   Cheddaring step. Drain off all the whey and allow the curds to knit together. Some whey will continue to come out of the curd. This should be removed regularly. The curd mass should be turned every 15 minutes and should be kept at between 34° and 36°C during this time. The total time for cheddaring will be about two hours. $pH \sim 5.3 - 5.4$

Note that the final pH plays an important role in the flavour, body, texture and colour of the cheese.

10. Break the curd with your hands and stir 25 to 30 gm of salt (for each 10 litres of milk) into them. Alternatively the curd may be chipped as seen below.

Possible Variation: At the salting stage blend in some herbs, onion or garlic.

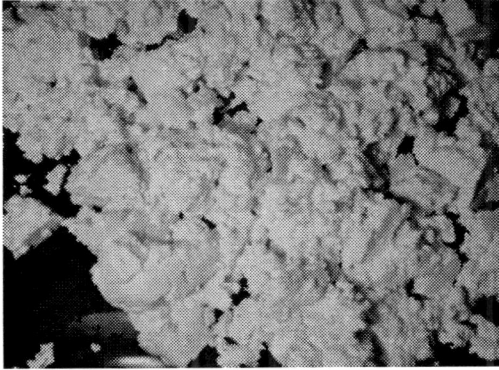

Curds broken for salting

Curd chipped for salting

11. Transfer the salted curds to a hoop, which is lined with cheesecloth.

12. Fold the cloth over neatly and place into the cheese press, and press for half an hour. Remove the cheese from the press, and rearrange the cloth to minimise the creases, then return the cheese to the press, maintaining the pressure until the following day. A weight of between ten and twenty kilograms should be enough.

13. Remove the cheese from the press and allow to dry. This may take up to four days. Placing the cheese on a cake cooling rack is often useful for drying the cheese.   pH ~ 5.1 – 5.3

14. The cheese must then be waxed before maturing. The waxing will be more successful if you apply a plastic cheese coating to the cheese first. Apply two coats of plastic cheese coating by brushing with a small paint brush. Allow the first coat to dry overnight before applying the second coat, then allow the second coat to dry before waxing.

15. To wax the cheese, the wax must be heated up until it is all molten. Holding one end dip the cheese into the wax for five to ten seconds. Remove and allow to dry. This should only take a couple of minutes. Once the wax is dry, turn it over and wax the other end. The wax temperature will determine the thickness of the coat. The optimum is between 90°C and 120°C.

16. Store the cheese for at least two months at 7 to 10°C before consumption. The flavour will be mild at this stage. If you want a stronger flavour, you will need to store it for several months.

CHESHIRE
***

Cheshire is an English cheese that is characterised by a high acid flavour and a flaky or crumbly body. It has a sharp acid flavour and is generally pale and even in colour. It originates from the Cheshire region and is now made in the surrounding counties, particularly Shropshire.

1. Pasteurise the milk and bring it to a setting temperature of 31°C. Add 250 mL of prepared Type A starter for each 10 litres, i.e. 2.5%. Mix in well.

2. Add rennet at a rate of two and a half mL for each 10 litres of milk. Dilute the rennet with at least 10 times its volume of cool boiled water, i.e. 25 mL of cool boiled water to each two and a half mL of rennet. Then pour the diluted rennet immediately into the milk, taking care to pour it over as much of the surface as possible, stirring all the time while pouring it in. Mix in well for no less than one minute and no more than three minutes. Maintain the setting temperature.

3. Allow the milk to set. This should take 30 to 35 minutes. pH ~ 6.6

4. Cut the curd into one cm cubes, then allow the curds and whey to stand for five minutes before stirring.

5. Over the next three hours, the curds and whey have to be stirred. Stir gently at first, and then more vigorously as the curd becomes firmer. As you start stirring, heat the curds and whey up to 34.5°C, taking 45 minutes.

6. Continue stirring intermittently until three hours has elapsed from the time of cutting the curd. All this stirring is to drive the whey out of the curds to make your cheese firmer. pH ~ 5.9 – 6.0

7. Texturing step. Drain off all the whey and allow the curds to knit together. Some whey will continue to come out of the curd. This should be removed regularly. The curd mass should be broken up every 30 minutes and should be kept at 34°C during this time. The total time for texturing will be about two hours. pH ~ 4.9 – 5.0

8. Cut the curd into small chip size and stir 20 gm of salt (for each 10 litres of milk used) into them.

9. Transfer the salted curds to a hoop and press for 48 hours at 20 to 22°C. A weight of between ten and twenty kilograms should be enough.

10. Remove the cheese from the press and allow to dry. This may take up to four days. A cake cooling rack is often useful for drying the cheese. pH ~ 4.7 – 4.9

11. The cheese may then be waxed before maturing to prevent it from drying out. The waxing will be more successful if you apply a plastic cheese coating to the cheese first. There are two types available, one with an anti mould agent and one without. The former is recommended. Apply two coats of plastic cheese coating by brushing with a small paint brush. Allow the first coat to dry overnight before applying the second coat, then allow the second coat to dry before waxing.

12. Store the cheese for at least two months at 7 to 10°C before consumption. It will be mild at this stage. If you want a stronger flavour, you will need to store it for several months.

CHEVRE
**

The word chevre is French for goat. Chevre is a soft goats milk cheese that is best soon after making. Coating the cheese with ash will soften the flavour.

1. Pasteurise the milk and bring it to a setting temperature of 25°C. Add 100 mL of prepared Type B starter for each 10 litres, i.e. 1%. Mix in well.

2. Add rennet at a rate of 0.5 mL for each 10 litres of milk. Dilute the rennet with at least 10 times its volume of cool boiled water, i.e. 5 mL of cool boiled water to each two and a half mL of rennet. Then pour the diluted rennet immediately into the milk, taking care to pour it over as much of the surface as possible, stirring all the time while pouring it in. Mix in well for no less than one minute and no more than three minutes. Maintain the setting temperature.

3. Allow the milk to set. This should take 4 hours

4. Gently ladle the curds to fill a plastic draining basket.

5. Continue filling the baskets as the whey drains from them until all the curds are in.

6. Wrap the cheese in a plastic film and refrigerate until consumed.

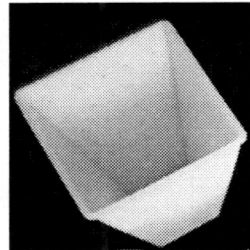

CHEVROTIN
**

The word chevre is French for goat. Chevrotin is a small soft goats milk cheese that is normally eaten within three to six weeks of making.

1. Pasteurise the milk and bring it to a setting temperature of 30 to 32°C. Add 150 mL of prepared Type B starter for each 10 litres, i.e. 1.5%. Mix in well. If you would like to have a rustic style Chevrotin with a cocktail of colours and aromas you will need to add other organisms. The possibles include yeasts, Geotrichum linens, Brevibacterium linens and micrococci. They will really enhance the flavour of the cheese.

2. Add rennet at a rate of two and a half mL for each 10 litres of milk. Dilute the rennet with at least 10 times its volume of cool boiled water, i.e. 25 mL of cool boiled water to each two and a half mL of rennet. Then pour the diluted rennet immediately into the milk, taking care to pour it over as much of the surface as possible, stirring all the time while pouring it in. Mix in well for no less than one minute and no more than three minutes. Maintain the setting temperature.

3. Allow the milk to set. This should take 40 to 45 minutes.

4. Cut the curd into small particles about the size of wheat grains. This is achieved by cutting and stirring followed by further cutting and so on until the curds are of uniform size.

5. Heat the curds and whey to 35°C whilst stirring.

6. Remove the curds from the whey using a cheesecloth and place into cloth lined hoops.

7. Press the curds overnight using a maximum weight of half to one kilogram.

8. After removing from the press, salt by placing in a brine of at least 20% salt for two to four hours.

9. Maturing takes place for one to two months at 12 to 15°C and at high humidity. The rind can be enhanced by washing the surfaces with either whey (one day old), or a weak brine solution at least three times per week. Most importantly here the cheese must be kept moist if the various surface organisms (added to the milk) are to grow and create their unique flavours.

CROTTIN
**

1. Pasteurise the milk and bring it to a setting temperature of 20-25°C. The room temperature is best to be the same.

2. Add 50 mL of prepared Type B starter for each 10 litres, i.e. 0.5%. Mix in well. White mould may be added if desired.

3. Add rennet at a rate of 0.5 mL per 10 litres of milk. Mix in well for no less than one minute and no more than three minutes. Maintain the setting temperature.

4. Allow the milk to set. This should take 18-24 hours.

    pH 4.6 - 4.7

5. Gently transfer the curds to a cloth bag for draining.

6. Remove curds and place into Crottin basket (see right) for moulding.

7. Salt with dry salt to 1.5% of cheese weight.

8. Store for up to one month at high humidity.

CREAMED GRANULAR COTTAGE CHEESE
***

Cottage cheese originated in Europe and was a soft fresh cheese made to use surplus milk. The American version of Cottage cheese is quite different to the European or Continental style. This recipe is for American style Cottage cheese and is known as granular Cottage. It is very popular in the USA, and is the most popular type of Cottage cheese eaten in Australia.

The milk must first have the fat removed as it is made from only skim milk. Hence it has a low fat content and is popular for people who are diet conscious. To enhance the cheese, light sour cream is blended with the curds, a process known as creaming. The overall fat content of the cheese is about the same as normal milk.

1.   Pasteurise the skim milk and bring it to a setting temperature of 25°C. Add 20 mL of prepared Type A starter for each 10 litres, i.e. 0.2%. Mix in well.

2.   Add rennet at a rate of one drop for each 10 litres of milk. Dilute the rennet with at least 10 times its volume of cool boiled water, i.e. 10 drops of cool boiled water to each drop of rennet. Then pour the diluted rennet immediately into the milk stirring all the time while pouring it in. Mix in well for no less than one minute and no more than three minutes. Maintain the setting temperature until step four.

3.   Allow the milk to set. This should take 14 to 18 hours.
$$pH \sim 4.6 - 4.7$$

4.   Cut the curd into one cm cubes, then allow the curds and whey to stand for five minutes.

5.   Pour onto the top of the curd enough warm water at 40 to 45°C to cover the curd by at least three centimetres.

6. Over the next two hours, the curds and whey have to be intermittently stirred, <u>extremely gently</u> for the first hour, five minutes stirring then five minutes rest, then stirring more vigorously as the curd becomes firmer. As you start stirring you should heat the curds and whey up to 55 to 60°C. The time for heating should take 90 to 120 minutes. The stirring should cease when the curds will pass the bounce test. This test requires you to take one of the curd particles and drop it onto the floor from shoulder height. If the curd bounces then it is ready for the next step.

7. Drain off the whey until the curds can just be seen above the whey, then add water at 30°C to the curds. The water quantity should be half the original skim milk volume. Stir for five minutes.

8. Drain off the water until the curds can just be seen above the water, then add 15°C water to the curds. The water quantity should be half the original skim milk volume. Stir for five minutes.

9. Drain off the water until the curds can just be seen above the water, then add 4°C water to the curds. The water quantity should be half the original skim milk volume. Stir for 10 minutes then drain all the water off over the next hour.

10. Stir in 20 grams of salt to 700 mL of light sour cream (for 10 litres of skim milk). Gently stir the salted cream into the curds before packing into tubs and refrigerating until use. The shelf life is two to three weeks.   pH ~ 4.9 – 5.2

Cottage cheese curd after the cooking process

FETTA
**

Fetta cheese originates from Greece, but is now made throughout the world. Denmark now makes more Fetta than any other cheese type, mostly for export. Fetta was traditionally made from sheep's milk, but now goat's and cow's milk are also used. With these milks it is not possible to create a Fetta that tastes the same as that made from sheep's milk. Fetta should have a flavour that is salty, sharp and acidic and if aged for more than two months slightly aromatic. A unique feature of Fetta cheese is that it spends its entire life in a brine solution. A feature of a well made Fetta cheese is that if rubbed into the palm of the hand, it should disappear like vanishing cream.

1.   Pasteurise the milk and bring it to a setting temperature of 32°C. Add 200 mL of prepared Type A starter for each 10 litres, i.e. 2%. Mix in well. Add one quarter of a teaspoon of either kid or lamb lipase powder to 50 mL of cool boiled water, mix to dissolve then pour into the milk. Mix in well. The Fetta can be made without the lipase, but adding it will give a more traditional flavour. Lipase is only needed when using cow's and sheep's milk.   pH ~ 6.6 – 6.7

2.   Add rennet at a rate of 1.6 mL for each 10 litres of milk. Dilute the rennet with at least 10 times its volume of cool boiled water, i.e. 16 mL of cool boiled water to each 1.6 mL of rennet. Then pour the diluted rennet immediately into the milk, taking care to pour it over as much of the surface as possible, stirring all the time while pouring it in. Mix in well for no less than one minute and no more than three minutes. Maintain the setting temperature.

3.	Allow the milk to set. This should take 60 to 90 minutes.

4.	Cut the curd into one cm cubes, then let stand.

5.	Stir the curd very gently, just to turn it over every hour until two hours have elapsed from the time of cutting.
pH ~ 6.1 – 6.3

6.	Drain off the whey, and pour the curds into perforated hoops on a cheesecloth lined draining tray.

7.	Invert the hoop after half an hour, and again in another half an hour. Then invert every two hours for the rest of the day. Alternatively place a light weight on top of the curd and no turning is needed.

8.	Let stand overnight.

9.	Next morning, remove the cheese from the hoops. Place the cheese into a container and cover with a 12-15% brine (brine pH 4.3). cheese pH ~ 4.6. The container should be slightly bigger than the cheese to avoid oversalting.

10.	Store the cheese in the brine solution in the refrigerator. It can be eaten immediately or stored for months in this way. If the cheese is too salty for your taste, remove it from the brine a couple of days before you want to eat it and place it in a container and cover it with milk. This will reduce the salt level and make it more palatable. pH ~ 4.3

A popular method of enhancing the presentation of Fetta cheese, is to place it into oil seasoned with herbs and spices. This can be done up to a week before eating.

FROMAGE BLANC
*

Fromage Blanc is a delicious tangy French cheese made from sour skim milk. You can make a similar cheese by mixing together Cottage cheese and natural yoghurt in approximately equal quantities.

1.   Take approximately 200 gm of cottage cheese and 200 mL of natural yoghurt, add 30 mL of lemon juice and half a teaspoon of sugar.
2.   Place into a blender and whip until thick and smooth.
3.   Cover and refrigerate overnight before eating.

FROMAGE FRAIS
**

Exactly what is meant by Fromage Frais depends on whether it is taken in the traditional context, whereby it simply means `fresh cheese', or the Australian current day commercialised version, which is a product that is a sweetened dessert. The following recipe gives a delicious dessert type of Fromage Frais. The base used is Quarg or Cream cheese.

•   Add 50 grams of sugar to 250 grams of your Quarg or Cream cheese and manually blend them together.
•   Slowly blend in 50 grams of pureed (cooked) apricots. (Use a whisk or whipping attachment on a food processor)
•   Cover and refrigerate overnight.

GJETOST

*

Gjetost cheese is made from goat's milk cheese whey, and is made in the same way as Mysost (see later recipe). Some goat's milk cream can be added to produce a cheese with a smoother consistency if desired.

GOAT'S MILK CHEESES

All cheese recipes can be used with goat's milk, although depending on the nature of the feed for the goat, the total solids level may be lower than that of cow's milk. Increasing the rennet quantity may be necessary, and setting times may be longer. See Chevre, Chevrotin, Crottin and Saint Maure for special goat's milk cheeses.

GORGONZOLA
****

This is one of Italy's most famous cheeses. There are references to this cheese as far back as 1000 AD. The cheese has its origins in Gorgonzola, a small village near Milan. It is a soft creamy "blue" cheese, although the real colour is closer to green. It is usually consumed between three and six months of age. This recipe involves a two curd process, using separate curd production for morning and evening milk.

1.   Pasteurise 10 litres of morning milk and bring it to a setting temperature of 32°C. Add 120 mL of prepared Type A starter for each 10 litres, i.e. 1.2%. Mix in well. Add five drops of blue mould suspension. Mix in well

2.   Add rennet at a rate of two and a half mL for each 10 litres of milk. Dilute the rennet with at least 10 times its volume of cool boiled water, i.e. 25 mL of cool boiled water to each two and a half mL of rennet. Then pour the diluted rennet immediately into the milk, taking care to pour it over as much of the surface as possible, stirring all the time while pouring it in. Mix in well for no less than one minute and no more than three minutes. Maintain the setting temperature.

3.   Allow the milk to set. This should take 30 to 35 minutes.

4.   Cut the curd into 12 mm cubes, then allow the curds and whey to stand for 10 minutes.

5.   Transfer the curds into a cheesecloth lined colander or a bag made of cheesecloth.

6.   Hang the curds over a sink to drain until the next day.

7.   Next morning make another batch of curd using steps 1 through to 5 as instructed above but this time using the evening milk.

8.   Instead of hanging the second curd overnight, allow it to hang for only one hour.

9.   Cut both lots of curds up into cubes of about two cm and place into separate bowls. Add four tablespoons of salt to each, then mix gently to distribute the salt evenly.

10.   The curds must then be placed into a hoop or mould. The warm curd must be placed in first, on the bottom and up the sides of the hoop. Use about half of the warm curd at this stage. Place the cold curd from the previous day into the centre and then completely enclose it with the rest of the warm curd.

11.   Turn the hoop over every 15 minutes for two hours then several times over the next three days.

12.   Rub the surfaces with salt each day for the next two days. It is best to have the salting area at 12°C during this time, but not essential.

13.   Wipe off any excess salt and punch holes through the cheese, about 20 cm apart. Use an ice pick or a size 10 knitting needle. It is best that the needle does not have a sharp point, as the holes will close up afterwards. The end should be partly blunted, but not fully blunt.

14.   Apply a solution of 4-6% salt containing Brevibacterium linens to the surface daily until the rind starts to turn a brownish colour.

15. The cheese should be stored at 10 to 12°C in a humid environment. It should be ready to eat in three to four months. If you prefer a milder flavoured cheese, the surface growth can be scraped off every three weeks.

Three matured Gorgonzola cheeses

GOUDA
***

This is a famous Dutch cheese, which takes its name from the town of Gouda, near Rotterdam in the Netherlands. Edam, also named after a town is made by the same method as Gouda. The main difference between the two cheeses being the shape, size and fat content. Gouda is a flat wheel of three to fifteen kilograms, whilst Edam is a cannonball shape, weighing one to three kilograms. Both of these cheeses have a special step during production, where the curds are pressed under the whey. This enables the cheese to be pressed in the absence of air, which is why the texture of the cheeses is very close with very few openings. During the maturing process small holes known as eyes may develop. This is caused by carbon dioxide gas being produced from one of the bacterial strains in the starter. It is normal for these holes to occur, but it is not a problem if none appear.
When making Edam, use this recipe and milk at about half of the original fat. See standardisation section.

1.   Pasteurise the milk and bring it to a setting temperature of 32°C. Add 150 mL of prepared Type B starter for each 10 litres, i.e. 1.5%. Mix in well.

2.   Add rennet at a rate of two and a half mL for each 10 litres of milk. Dilute the rennet with at least 10 times its volume of cool boiled water, i.e. 25 mL of cool boiled water to each two and a half mL of rennet. Then pour the diluted rennet immediately into the milk, taking care to pour it over as much of the surface as possible, stirring all the time while pouring it in. Mix in well for no less than one minute and no more than three minutes. Maintain the setting temperature.

3. Allow the milk to set. This should take 30 to 40 minutes.

4. Cut the curd into 13 mm cubes, then let stand for five minutes before stirring.

5. Stir the curd gently and regularly over the next 40 to 50 minutes, being sure to maintain the temperature at about 32°C.

6. Allow the curd to settle and remove whey to half the original milk level.

7. Cooking: Slowly add two litres of boiled water cooled to 60°C. Add the water over 15 minutes, and at the end of this time the temperature of the curd/whey mixture should have just reached 38°C. If the temperature does not reach 38°C, you will need to heat by one of the indirect methods (see section on cooking the curds and whey). Stir frequently during this time.

8. Stir the curd frequently over the next 30 mins. During the cooking process the curd becomes firmer, thus the intensity of the stirring can be increased without shattering the curd.

9. Take off the whey to the point where it just covers the surface of the curd. Collect some of this whey as you may need to use it to cover the curd in the next step.

10. Place the hoop that you are going to use to press and shape the cheese, into a slightly larger vessel. Put the curds and whey/water mix into the hoop, and press the curd down slightly while it is under the whey. At this stage, the whey must be covering the curd completely. If it doesn't cover the curd, you can add some of the spare whey that you collected. Press the curd for 15 minutes. A large two litre jar filled with

water will have enough weight for pressing at this stage.

11.   Drain the whey from the outside vessel, while being sure not to disturb the curd in the hoop. Press the curd overnight. A weight of 10 kilograms should be sufficient.

12.   Next morning, place the cheese in a cold 20% brine solution (i.e.. 200 gm salt to each 800 mL of cool boiled water).   pH ~ 5.3 – 5.4 The amount of time the cheese requires in the brine depends on the size of the cheese. As a guide, 250 gram requires about two hours
500 gram requires about four hours
1 kilogram requires about eight hours.

13.   Remove the cheese from the brine, and allow to dry. It will take a minimum of one day to dry, perhaps longer.

14.   When dry, apply two coats of plastic cheese coating, followed by a coat of wax.

15.   Store at 10 to 15°C for two to six months, turning the cheese two to three times per week.

HALLOUMI
**

Halloumi is a hard cheese made predominantly in Cyprus, from sheep's or goat's milk. It can be made from cow's milk quite successfully.

1.   Heat the milk to between 32 and 34°C.

2.   Add rennet at a rate of two and a half mL for each 10 litres of milk, diluting it with at least 10 times its volume of cool boiled water, i.e. 25 mL of cool boiled water to each two and a half mL of rennet. Then pour the diluted rennet immediately into the milk, taking care to pour it over as much of the surface as possible, stirring all the time while pouring it in. Mix in well for no less than one minute and no more than three minutes. Maintain the setting temperature.

3.   Allow the milk to set. This should take 40 minutes, and should be a firm set.

4.   Cut the curd into 20 to 40 mm cubes, then let stand for five minutes before stirring gently and heating to 40°C over 20 minutes.

5.   Allow the curd to settle to the bottom and form into one solid mass.

6.   Remove all of the whey, then press down on the curd to help it knit together, before placing it into a cheese cloth.

7.   Press the curd in the cloth by hand by tightening the cloth around the curds. Finally weights may be placed on top until it is firm enough for your liking.

8.   Heat the whey to 90°C and collect with a ladle any curd that rises to the surface. Bring the whey to boiling point.

9. Remove the cheese from the cloth, cut it into blocks of size 50 x 100 x 150 mm and place into the hot whey. The curd will sink to the bottom. After 45 minutes and up to 90 minutes the curd will start to float. When all of the curd has risen to the surface, wait a further 15 minutes, then remove and place on a wooden rack.

10. After twenty minutes the cheese is ready for salting. Sprinkle salt on the cheese and leave them until they are cold. A variation of Halloumi can be made at this stage by placing some mint on the top of the cheese, then folding it in half with the mint in the middle, before salting and cooling.

11. Place the cheese into a brine solution. The brine can be made by adding 300 gm of salt to each litre of cool boiled water.

12. Store the cheese in the brine solution in the refrigerator. It can be eaten immediately or stored for months in this way.

HAVARTI
***

Havarti is a cheese that originates from Denmark, and is a copy of the Tilsit cheese that is made in the North of Germany. It is made in two versions, standard, and a creamy version which has extra cream added to the milk. To be a true Havarti, the cheese should have its rind washed regularly during the maturing process, a step which helps create the typical flavour.

1. Pasteurise the milk and bring it to a setting temperature of 32°C. Add 150 mL of prepared Type B starter for each 10 litres, i.e. 1.5%. Mix in well.

2. Add rennet at a rate of two and a half mL for each 10 litres of milk, diluting it with at least 10 times its volume of cool boiled water, i.e. 25 mL of cool boiled water to each two and a half mL of rennet. Then pour the diluted rennet immediately into the milk, taking care to pour it over as much of the surface as possible, stirring all the time while pouring it in. Mix in well for no less than one minute and no more than three minutes. Maintain the setting temperature.

3. Allow the milk to set. This should take 40 minutes. This should be a firm set.

4. Cut the curd into 10 mm cubes, then allow the curds and whey to stand for five minutes before stirring gently for 20 minutes.

5. Drain off one third of the whey. Slowly add two litres of pasteurised water (for each 10 litres of milk used) at about 60°C. The final temperature of the whey should be 38°C.

6. Stir intermittently for 70 minutes, while keeping the curd and whey at 38°C.

107

7. After this 70 minutes has elapsed, drain half of the whey off, then add cold water to lower the temperature to 28°C.

8. Five minutes, later drain off all of the whey and briefly stir the curds to ensure they are free of whey.

9. Pour the curd into hoops. Turn hoops over frequently.

10. After about two hours from the time of hooping, cool the cheese in cold water at 8 to 10°C. Leave the cheese in this cool water for two hours. If you are making bigger cheese you will need to cool for longer, and if you are making smaller cheese you will need to cool for less time. The minimum time in the water is one hour, while the maximum time is three hours.

11. Place the cheese in a 20% brine solution (200 gm salt to 800 mL of cool boiled water) at 8 to 10°C for three to four hours.

12. Drain on a cake rack at room temperature, for one to three days, until the surface of the cheese is completely dry.   pH ~ 5.2 – 5.3

13. Apply two coats of plastic cheese coating and then one coat of wax.

14. Store at 10°C turning twice weekly. The cheese will be ready to eat in six to 10 weeks.

The cheese should have an open texture. If the texture of the cheese is not open enough, you may need to either keep the curds and whey stirring for longer before wheying off, or give the curds some more dry-stirring before hooping, at step 8.

The more traditional method of Havarti manufacture uses the same steps up until step 12. A technique of washing the cheese surface every second day, is then used instead of applying plastic cheese coating and wax. The solution used to wash the cheese is a mild brine solution, containing a special bacteria known as Brevibacterium linens. After three to four weeks, the rind develops a sticky surface growth(see below) that may vary from orange to reddish in colour. During this period the cheese should be stored on a wooden shelf and turned after each wash. This method results in a cheese different in flavour than the previous method. The cheese will have an aroma that you may consider to be less than pleasant, but this is normal, and is not reflected in the flavour of the cheese.

Note the surfaces on these traditional Havarti cheeses

LEMON CHEESE
*

This cheese is very simple and can be made with any quantity of milk. For each litre of milk, add the juice of a small lemon. Then simply heat the milk up until the milk curdles and the whey becomes a green to yellow colour. Remove from the heat and collect the curd in a cheesecloth lined colander. Some grated lemon rind can be added to the curd for added flavour. Tie the corners of the cloth tightly on the cheese and allow the whey to drain out for a couple of hours. Refrigerate and use fresh.

MASCARPONE
*

Mascarpone is an acidified cream cheese which is traditionally made in Italy throughout the autumn and winter months.

1.   Heat some pure or rich cream to 90°C in a saucepan.

2.   Acidify it with either vinegar, tartaric acid, lemon juice. Just stir in enough to thicken (curdle) the cream.

3.   Transfer the mixture into a container and refrigerate.

It can be eaten fresh and has a clean buttery flavour.

MOZZARELLA
***

Mozzarella cheese is a unique Italian cheese that is mostly used as a pizza topping. It obtains its stretchy, stringy texture during a step where the curds are immersed into a bath of hot water, and are worked until smooth. It is a cheese that has a mild flavour and should be eaten fresh. The same process is used to make Provolone, except that some lipase enzyme is added to the milk at the same time as the starter. A Provolone is matured for a few months, during which time it becomes firmer and develops a stronger flavour.

You will need to have a cold 20% brine solution ready before making this cheese.

1.   Pasteurise the milk and bring it to a setting temperature of 36°C. Add 150 mL of prepared Type C starter for each 10 litres, i.e. 1.5%. Mix in well.

2.   Add rennet at a rate of two and a half mL for each 10 litres of milk, diluting it with at least 10 times its volume of cool boiled water, i.e. 25 mL of cool boiled water to each two and a half mL of rennet. Then pour the diluted rennet immediately into the milk, taking care to pour it over as much of the surface as possible, stirring all the time while pouring it in. Mix in well for no less than one minute and no more than three minutes. Maintain the setting temperature.

3.   Allow the milk to set. This should take 30 to 35 minutes.   pH ~ 6.5 – 6.6

4.   Cut the curd into 13 mm cubes, then let stand for five minutes before stirring.

5. Stir gently and warm the curds and whey to 41°C over the next 40 minutes.

6. Stir intermittently over the next one to one and a half hours.

7. Drain off the whey.   pH ~ 6.1 – 6.2

8. Keep the curd warm at 42 to 43°C. This will cause the curd to fuse together into a single mass. Turn the mass of curd over every 15 minutes to keep it warm. Whey will continue to come out of the curd and should be drained off at regular intervals.

9. Stretch test. About one hour after removing the whey, take a small thin slice of curd and place into water at 70°C. When the curd is warmed through, remove it from the water and work it well with the fingers. Using the thumb and forefinger of both hands, stretch the curd until it breaks. If the curd is ready to stretch it will be very elastic. If the curd is brittle, it is not ready to stretch. (If you have pH test strips, the curd will stretch when the pH is between 5.0 and 5.4).

10. If the curd is not ready to stretch, continue to keep the curd warm (as in step 8) and do a stretch test (as in step 9) every 15 minutes until it is ready.

11. When curd is ready to stretch, cut the curd into very thin slices, and place into hot water at 70°C. You may find it easier to place the sliced curd into a colander, as you can easily raise and lower the curd in and out of the water. To stretch the curd, first work the curd with your hands until there is no openness in the curd and it is quite elastic. You will need to pick up the curd in your hands and pull it apart. At this stage it should not break, but should stretch out. You should then fold it together and repeat the stretching, until all the curd has been well worked.

12. Using such a high water temperature, you will need to wear thick rubber gloves while working the curd or you will burn your hands! You can use two wooden spoons or ladles instead, but we find using our hands gives a better feel of the texture of the curd.

13. Shape the cheese into the desired form. You can shape into a ball with your hands, or you can put it into a container in the shape of the cheese you want.

14. Place the cheese into ice cold pasteurised water, to set the shape that you have just made. Leave the cheese in the water for one hour. If making a bigger cheese, leave it in the water for longer, and if making a smaller cheese, it will need less time in the water.

15. Place the cheese into the cold brine solution for about four hours, for cheese made from 10 litres of milk. If you are making a bigger mozzarella, you will need to leave it longer, and if you are making a smaller mozzarella, you will need to leave it for less time. The cheese is ready to eat immediately, or can be stored for up to a month in the fridge.

MYSOST
*

Mysost originates from the Scandinavian countries of Norway and Sweden. It is a whey based cheese, which does not really resemble what we think of as cheese. It is a product with a combination of sweet, cooked and salty flavours. The process merely involves boiling the whey for several hours until it becomes a syrup. The syrup hardens on cooling and can be used as a spread.

1.   Take the whey from your cheesemaking, place it in a large pot or saucepan. Place on a slow combustion stove and bring to the boil. You must be careful not to boil too quickly or it will boil over. After a short while you will notice some white curds appear. These can be removed and placed into a bowl for adding back later.

2.   Allow the whey to simmer for several hours until it starts to thicken. The whey should then be stirred occasionally until it reaches the consistency of a thin porridge. At this point the curds removed earlier in the day can be stirred back in. It can then be boiled longer, or cooled. If you want the product to be soft and spreadable, cool immediately. If you want a firmer Mysost, boil away until it reaches the consistency of a thick porridge before cooling. As a guide, you can remove a sample with a spoon and cool it quickly. This will give you an indication of the consistency you can expect in the final product.

PARMESAN
****

This cheese is by far the most important cheese in Italy. It is also known as a Grana (meaning granular) cheese and has many variations such as Parmigiano, Reggiano, Sardo and Padano. They are all similar and vary mainly in size, shape and weight. These Grana varieties are all very hard and as such are excellent grating cheeses. Being so hard comes about from the fact that they have low moisture levels, which is also the reason that they keep for so long. A Grana cheese is best eaten after one year, but will last for several years.

Before making Parmesan, the milk needs to be reduced to about half its normal fat content. This can be done by mixing equal parts of whole milk and skim milk together, or allowing the milk to settle before removing half of the cream from the top. See standardisation section for more details.

1.   Pasteurise the milk and bring it to a setting temperature of 35°C. Add 100 mL of prepared Type C cheese starter for each 10 litres, i.e. 1.0%. Mix in well. Allow to sit for 15 minutes. Add one quarter of a teaspoon of either kid or lamb lipase powder to 20 mL of cool boiled water, mix to dissolve, then pour into the milk. Mix in well. The Parmesan cheese can be made without the lipase, but adding it will give a more traditional flavour.

2.   Add rennet at a rate of two and a half mL for each 10 litres of milk, diluting it with at least 10 times its volume of cool boiled water, i.e. 25 mL of cool boiled water to each two and a half mL of rennet. Then pour the diluted rennet immediately into the milk, taking care to pour it over as much of the surface as possible, stirring all the time while pouring it in. Mix in well for no less

than one minute and no more than three minutes. Maintain the setting temperature.

3. Allow the milk to set. This should take 20 to 30 minutes.

4. Cut the curd into four mm cubes, this is about wheat grain size, then let stand for five minutes before stirring.

5. Stir for 10 to 15 minutes, while maintaining the setting temperature.

6. Increase temperature slowly to 42°C, taking half an hour to reach the temperature, being sure to stir regularly. Hold this temperature for 15 minutes, stirring regularly.

7. Increase temperature to between 51 and 54°C, taking half an hour to reach the temperature, stirring regularly.

8. Leave for 30 minutes, letting curd settle.

9. Run off whey, and place curd into a cheesecloth lined hoop.

10. Press the cheese overnight.

11. Make up a 20% brine solution (200 gm salt to each 800 mL of cool boiled water). Place cheese into brine for 8 to 12 hours.

12. Put the cheese on rack to dry. When it is dry, apply two coats of plastic cheese coating before waxing.

13. Store the cheese at between 10 and 15°C. It normally takes about one year to mature. It can be consumed before this time, but it will not have the characteristic strong flavour.

PEPATO
***

This Italian cheese can be made from cow's, sheep's or goat's milk. Traditionally it is made from sheep milk and is known as Pecorino Pepato. Before making Pepato, the milk needs to be reduced in fat. This can be done by mixing whole milk and skim milk together in a ratio of 3:2, or allowing the milk to settle before removing one third of the cream from the top. This will result in a harder cheeses that can be used for grating purposes. The unique flavour of Pepato comes from both the peppercorns the lipase enzyme that is added to the milk.

1. Pasteurise the milk and bring it to a setting temperature of 37°C. Add 200 mL of prepared Type C cheese starter for each 10 litres, i.e. 2.0%. Mix in well. Allow to sit for 15 minutes. Add one quarter of a teaspoon of either kid or lamb lipase powder to 50 mL of cool boiled water mix to dissolve then pour into the milk. Mix in well. pH ~6.6-6.7

2. Add rennet at a rate of two and a half mL for each 10 litres of milk, diluting it with at least 10 times its volume of cool boiled water, i.e. 25 mL of cool boiled water to each two and a half mL of rennet. Then pour the diluted rennet immediately into the milk, taking care to pour it over as much of the surface as possible, stirring all the time while pouring it in. Mix in well for no less than one minute and no more than three minutes. Maintain the setting temperature.

3. Allow the milk to set. This should take 30 minutes.

4. At this stage bring to the boil a small quantity of water then add dried black peppercorns. Boil for 20 minutes.

Use about 10-20 grams of peppercorns for 10 litres of milk.

5.   Cut the curd into six mm cubes, then let stand for five minutes before stirring.

6.   Heat slowly to 45°C over one hour, whilst stirring. Continue stirring for another 30 minutes.

7.   Drain off all the whey, and transfer curds to a cheesecloth lined hoop.   pH ~6.1-6.3

8.   Press the cheese for one hour, remove it, then turn it over end for end, before readjusting the cheesecloth to remove any wrinkles.

9.   Press the cheese overnight. A weight of between ten and twenty kilograms should be enough.

10.   Place the cheese in a 26% (saturated) brine at 10 to 15°C for eight hours.   pH ~5.0-5.1

11.   Allow cheese to dry on a wire rack. This could take up to two days.

12.   Apply one coat of plastic cheese coat, and allow to dry. Repeat for a second coat.

13.   Apply a coat of wax. Store the cheese at 10 to 15°C. The cheese can be eaten from two months old, or stored for up to 12 months.

QUARG/CREAM CHEESE
*

Quarg or Quark originates from Germany and is very easy to make. There are two versions, one made with skim milk and the other with whole milk. They are soft white cheeses eaten soon after making. Cream cheese can also be made using the same method. This recipe produces excellent quality cheese when using UHT milk. UHT goat's milk is no exception. The cheeses made from UHT milk have a finer texture and delicious flavour.

By using skim milk instead of whole milk you can make a low fat Quarg. By adding 200 mL of cream to two litres of wholemilk you can make a cream cheese. Both of these variations use the same basic procedure, but you will find that the skim milk version drains quicker, and the Cream cheese version drains slower than when using wholemilk. This is an excellent base for fruit cheese, or cheesecake.

1. Add one tenth of a teaspoon of DVS Type A starter powder or two teaspoons of liquid Type A starter to two litres of pasteurised or UHT milk. Mix in well. Warm the milk to 20°C and leave overnight in a warm place. The milk will curdle in 12 to 24 hours.

    pH ~ 4.4 – 4.6

2. Pour the curd into a cheese cloth lined colander, then hang the cloth to drain until sufficient whey has drained out. Drainage time can be quite variable depending on temperature, type of cheese cloth and fat content of the milk. The drainage should be stopped when the curd is dry enough for your liking. Drainage can be sped up by manually working the bag at intervals with your hands, or by placing a weight on the bag instead of hanging it.

3.   Remove cheese from cloth and put into a bowl. Add salt to taste and mix well. Cover and refrigerate until ready to use.

Chocolate cheese log

A delicious chocolate cheese log can be made from the Quarg or cream cheese. Take 200 grams of cheese, place in a bowl then add one third cup of good quality chocolate hail. Mix cheese and chocolate together with a spoon. Remove from the bowl, shape into a log then roll in coconut. Wrap in plastic wrap before chilling in a refrigerator for 24 hours before eating. If the cheese log is too moist then you will need to make the cheese a little drier next time.

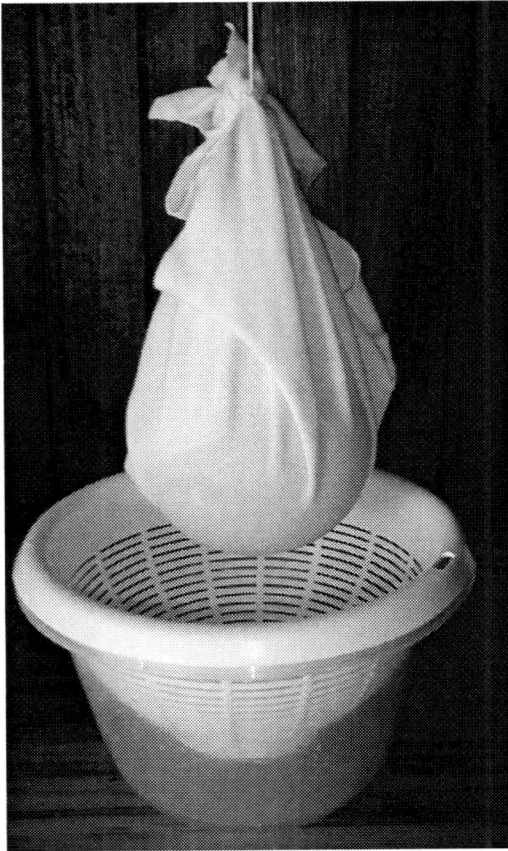

Quarg draining in a cheesecloth hung over a colander.

Herb cheese log

The same process can be applied for a herb cheese log by adding herbs to the cheese, mixing, then shaping to a log or ball and rolling in either crushed pepper or a coating of your choice.

Fruit cheese log

A fruit log can be made by adding 100 grams of dried fruit medley to 250 grams of cheese, mixing, then shaping to a log or a ball and coating with poppy seed. With fruit cheese the fruit tends to absorb the water from the cheese, hence the cheese needs to be slightly more moist than for the other two cheese logs, or the cheese will crumble. An alternative is to soak the fruit in your favourite liqueur prior to adding to the cheese.

RED LEICESTER
***

Proceed with the method as for cheddar cheese except that the milk is left for 30 minutes after the starter has been added but before the rennet is added. Shortly after the starter is added annatto cheese colour is added at the rate of 0.5 mL per litre of milk used.

RICOTTA
*

Another Italian cheese, it is unique in that it is made from the whey that separates from the curd during cheesemaking. Use only fresh whey when making Ricotta cheese. This is the whey obtained from a cheese that uses rennet to set the curd. Whey from cheese such as Quarg is not suitable, because it is acidic or sour. The whey should not be held for any longer than one hour before making Ricotta, or it will continue to acidify and will be no longer suitable. It is important that the pot used for Ricotta is deeper than it is wide, preferably twice as high as wide, to ensure a good yield of Ricotta.

1. Put five litres of fresh whey into a saucepan. Heat on a hot plate to 60°C, stirring regularly to avoid burning any milk solids. pH ~ 6.3 – 6.5

2. Add two cups of whole milk and a teaspoon of salt after the whey reaches 60°C.

3. Continue to heat to 95°C, stirring constantly.

4. Add about 20 to 40 mL of white vinegar, which has been diluted in 200 mLs of water, and stir in quickly for about 10 seconds. The actual amount of vinegar to add will depend on the strength of the vinegar and the

freshness of the whey. Stir quickly while gradually adding the vinegar. At the first sign of specks appearing in the whey, stop adding the vinegar.

pH ~ 5.6 – 5.8

5.   Let stand for about five minutes. Gently scoop off the layer of curd that rises to the surface, and lay it into a cheesecloth lined colander. This is your Ricotta. It can be used immediately or stored in the fridge and used within a week..

•   If the whey is too acidic, precipitation may start to occur before the desired temperature is reached. Depending upon the stage at which the precipitation occurs, you will get either a reduced amount of Ricotta, or no Ricotta at all.

Ricotta made from whey, floating to the surface

RICOTTA (WHOLE MILK)
*

Believe it or not, true Ricotta is made from whole milk using the same method as for the whey based Ricotta. The flavour, however, is not the same as for the Ricotta made from whey. It will give almost two kilograms of cheese from 10 litres of milk. Goat's, sheep's and cow's milk can all be used successfully.

1.   Heat 10 litres of milk to 90 to 95°C in a pot on a hot plate, stirring continually. Smaller quantities can be microwaved.

2.   As the milk reaches temperature, quickly stir in half a cup of vinegar. The milk will coagulate.

3.   Transfer the curds to a cheesecloth lined colander, draining for five minutes. Transfer the curds to a mixing bowl and blend in a little salt and butter if desired.

4.   Refrigerate until ready to use.

Note the gentle layering of the Ricotta from ladle to basket.

ROMANO
***

This Italian cheese can be made from cow's, sheep's or goat's milk. The cow's milk version is traditionally known as Vacchino Romano, the sheep's milk version Pecorino Romano and the goat's version Caprino Romano. Before making Romano, the milk needs to be reduced in fat. This can be done by mixing whole milk and skim milk together in a ratio of 3:2, or allowing the milk to settle before removing one third of the cream from the top. This will result in a harder cheeses that can be used for grating purposes. The unique flavour of Romano comes from the lipase enzyme that is added to the milk.

1. Pasteurise the milk and bring it to a setting temperature of 37°C. Add 200 mL of prepared Type C cheese starter for each 10 litres, i.e. 2.0%. Mix in well. Allow to sit for 15 minutes. Add one quarter of a teaspoon of either kid or lamb lipase powder to 50 mL of cool boiled water mix to dissolve then pour into the milk. Mix in well.   pH ~ 6.6 – 6.7

2. Add rennet at a rate of two and a half mL for each 10 litres of milk, diluting it with at least 10 times its volume of cool boiled water, i.e. 25 mL of cool boiled water to each two and a half mL of rennet. Then pour the diluted rennet immediately into the milk, taking care to pour it over as much of the surface as possible, stirring all the time while pouring it in. Mix in well for no less than one minute and no more than three minutes. Maintain the setting temperature.

3. Allow the milk to set. This should take 30 minutes.

4. Cut the curd into six mm cubes, then let stand for five minutes before stirring.

5. Heat slowly to 45°C over one hour, whilst stirring. Continue stirring for another 30 minutes.

6. Drain off all the whey, and transfer curds to a cheesecloth lined hoop.   pH ~ 6.1 – 6.2

7. Press the cheese for one hour, remove it, then turn it over end for end, before readjusting the cheesecloth to remove any wrinkles.

8. Press the cheese overnight. A weight of between ten and twenty kilograms should be enough.
                                                    pH ~5.0 – 5.1

9. Place the cheese in a 26% (saturated) brine at 10 to 15°C for eight hours.

10. Allow cheese to dry on a wire rack. This could take up to two days.

11. Apply one coat of plastic cheese coat, and allow to dry. Repeat for a second coat.

12. Apply a coat of wax. Store the cheese at 10 to 15°C. The cheese can be eaten from two months old, or stored for up to 12 months.

SAINT MAURE
**

A delicate soft goat's milk cheese that is ripened with a white mould. It is traditionally made in Touraine, France, during the months of April through November. It is usually cylindrical in shape, weighing about 200 grams.

1. Pasteurise the milk and bring it to a setting temperature of 22 to 23°C. Add 50 mL of prepared Type A starter for each 10 litres, i.e. 0.5%. Mix in well.

2. Add rennet at a rate of one mL for each 10 litres of milk, diluting it with at least 10 times its volume of cool boiled water, i.e. 10 mL of cool boiled water to each one mL of rennet. Then pour the diluted rennet immediately into the milk, taking care to pour it over as much of the surface as possible, stirring all the time while pouring it in. Mix in well for no less than one minute and no more than three minutes. Maintain the setting temperature.

3. Allow the milk to set. This should take 18 to 24 hours.

4. Ladle the curd into the drainage hoops, taking care not to damage or break up the curds.

5. Leave the curd to drain for the next 24 to 48 hours. There is no need to turn the hoops during drainage, although if you prefer a more regular shaped cheese occasional turning will help.

6. Remove the cheese from the hoops. Be very careful not to break them as they are quite fragile.

7. Sprinkle dry salt onto the surfaces of the cheeses. Use about 50 grams of salt for each kilogram of cheese, but this can be varied to suit your taste.

8.   Store the cheese for two weeks, or alternatively they may be sprayed with a suspension of white mould spores and then stored at 11 to 12°C and high humidity (90 to 95%), until covered with mould. The cheese is best eaten from two and five weeks after making.

SOUR CREAM CHEESE
*

Make sour cream as per the instructions in the sour cream recipe. Transfer the sour cream to a cheesecloth and hang for 24 hours to drain. Then place the cream into a bowl, adding salt, and herbs (either caraway or cumin can be added). A little grated parmesan may also be added to enhance the flavour. Shape into block or fill into a ceramic pot. Refrigerate until used.

STILTON
****

Stilton is a famous English blue cheese. It is made with milk with a little higher than the normal fat level. Normal milk may be used, but if you would like to make it a little creamier, then refer to the section on standardisation.

1. Pasteurise the milk and bring it to a setting temperature of 31°C. Add 70 mL of prepared Type A starter for each 10 litres, i.e. 0.7%. Mix in well and add two drops of blue mould suspension. Mix in well again.

2. Leave the milk to ripen with the starter for 30 minutes. Maintain at 31°C.

3. Add rennet at a rate of one mL for each 10 litres of milk. Dilute the rennet with at least 10 times its volume of cool boiled water, i.e. 10 mL of cool boiled water to each one mL of rennet. Then pour the diluted rennet immediately into the milk, taking care to pour it over as much of the surface as possible, stirring all the time while pouring it in. Mix in well for no less than one minute and no more than three minutes. Maintain the setting temperature until step 8.

4. Allow the milk to set. This should take about 100 minutes.

5. Cut the curd into 12 mm cubes. Allow the curds and whey to sit for 10 minutes.

6. Transfer the curds gently into a cheesecloth lined colander. The colander should sit in a pot or pan so that all the whey can collect and cover the cheese. This is to collect the curds together and to keep them warm.

7. Allow the curds to rest in the whey for 100 minutes.

8. Gather the cheese cloth so that it forms a bag, then hang it to drain until only a little whey is dripping out.

9. Take the bag down and place onto a draining tray. Place a weight of four to five kilograms on top of the bag to press overnight.

10. Next morning break the curd into walnut size pieces. Sprinkle three tablespoon of salt over the curds and gently mix in.

11. Fold the cloth over neatly and place into the cheese press, and press for half an hour. Remove the cheese from the press, rearrange the cloth to minimise the creases, then return the cheese to the press, maintaining the pressure for three days.

12. Next day rub a strong brine solution (25%) on the cheese surfaces and leave it for 24 hours.

$$pH \sim 4.6 - 4.7$$

13. Punch holes into cheese, about 20 mm apart. Use an ice pick or a size 10 knitting needle. It is best that the needle does not have a sharp point, as the holes will close up afterwards. The end should be partly blunted, but not fully blunt or the cheese may break apart.

14. Mature the cheese at 12 to 14°C for two weeks, then at 10 to 12°C for three months. The humidity should be kept high during this time. Turn the cheese daily. You may wish to leave the mould and slimy growth on the cheese or scrape it off at regular intervals. If left on the surface, a strong pungent aroma will probably result.

Stilton curd now in the hoop.

SWISS
****

In Switzerland there is no cheese named Swiss cheese. There are four Swiss cheese types, known as Emmenthaler, Appenzaler, Gruyere and Sbrinz. Each has its own traditional shape and size, and region of manufacture. Over 1000 village based factories operate in Switzerland, each making only a handful of cheeses each day.

Don't be under any illusion, this is not an easy cheese to make. The main difficulty is obtaining the correct body and eye formation. This version is known as a sweet Swiss cheese. It is simpler to make than a traditional Swiss cheese such as Emmenthaler, which can weigh between 80 and 120 kilogram.

1.   Pasteurise the milk and bring it to a setting temperature of 32°C. Add 100 mL of prepared Type B starter for each 10 litres, i.e. 1.0%. Mix in well. Add a tiny quantity of Propionic acid bacteria. This will help produce the eyes.   pH ~ 6.6 – 6.7

2.   Add rennet at a rate of two and a half mL for each 10 litres of milk, diluting it with at least 10 times its volume of cool boiled water, i.e. 25 mL of cool boiled water to each two and a half mL of rennet. Then pour the diluted rennet immediately into the milk, taking care to pour it over as much of the surface as possible, stirring all the time while pouring it in. Mix in well for no less than one minute and no more than three minutes. Maintain the setting temperature.

3.   Allow the milk to set. This should take 30 to 40 minutes.

4.   Cut the curd into 13 mm cubes, then let stand for five minutes.

5. Stir the curd gently and regularly over the next 40 to 50 minutes, being sure to maintain the temperature at about 32°C.

6. Slowly take off half the whey

7. Cooking: Slowly add two litres of boiled water cooled to 60°C. Add the water over 15 minutes, so that at the end of this time the temperature of the curd/whey mixture should have just reached 38°C. Stir frequently during this time.

8. Stir the curd frequently over the next 30 mins. During the cooking process the curd becomes firmer, and so the intensity of the stirring can increase now without shattering the curd.

9. Take off the whey to the point where it just covers the surface of the curd.   6.3 – 6.4

10. Place the hoop that you are going to use into a slightly larger vessel. Put the curds and whey/water mix into the hoop, and press the curd down slightly while it is under the whey. At this stage the whey must cover the curd completely. Press the curd for 15 minutes. A weight of four kilogram is sufficient.

11. Drain the whey from the outside vessel, while being sure not to disturb the curd in the hoop too much.

12. Press the curd overnight.

13. Next morning, place the cheese in a cool (10 to 15°C) 20% brine solution. The amount of time the cheese requires in the brine depends on the size of the cheese. About one day is enough for a five kilogram cheese.   pH ~ 5.2 – 5.3

14. Remove the cheese from the brine, and store at 18 to 20°C for four to six weeks, turning the cheese three times per week. If the cheese appears to be drying

out it can be washed with a mild brine solution or fresh whey each time it is turned.

15. After this storage time the cheese can be plastic coated and waxed.

16. Store the cheese at 10°C until it is three months old.

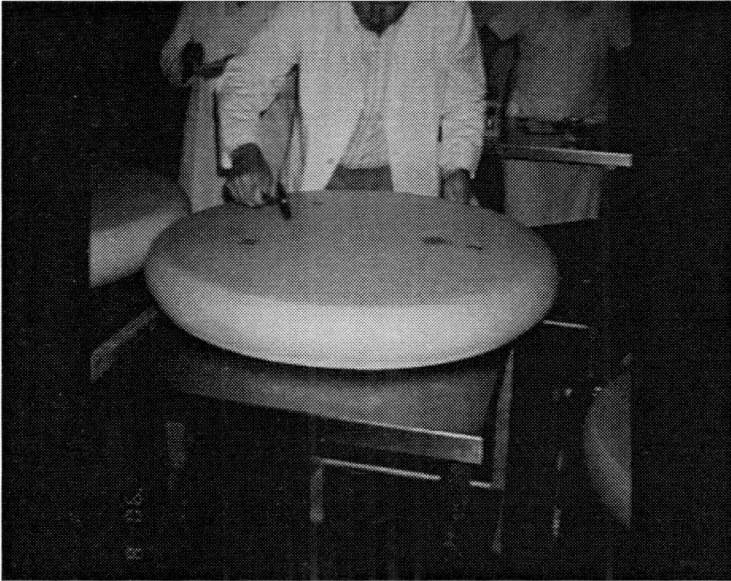

Grading of a traditional Emmenthaler relies on tapping the cheese rather than taste. The eyes provide a hollow sound.

WASHED RIND – SOFT
****

1. Pasteurise the milk and bring it to 40 to 42°C. Add 200 mL of prepared Type E starter for each 10 litres, i.e. 2%. Alternatively use an equal mixture of Type B and E. Mix both in well. Leave for 45 minutes.
    pH ~ 6.6 – 6.7

2. Add rennet at a rate of two and a half mL for each 10 litres of milk. Dilute the rennet with at least 10 times its volume of cool boiled water, i.e. 25 mL of cool boiled water to each two and a half mL of rennet. Then pour the diluted rennet immediately into the milk, taking care to pour it over as much of the surface as possible, stirring all the time while pouring it in. Mix in well for no less than one minute and no more than three minutes. Maintain setting temperature until step 11.

3. Allow the milk to set. This should take 30 to 35 minutes.   pH ~ 6.4 – 6.5

4. Cut the curd into two cm cubes .

5. Allow to sit for 30 minutes.

6. Turn all the curd over gently, for three minutes

7. Allow to sit for 30 minutes.

8. Turn all the curd over gently again as in step 6.

9. Allow to sit for 30 minutes.

10. Turn all the curd over gently again as in step 6.

11. Allow to sit for 30 minutes.

12. Drain off half of the whey and pour the remaining curd into hoops. The hoops should be placed onto a draining tray lined with cheesecloth.   pH ~ 6.0 – 6.2

13. Invert the hoops after 10 minutes and again after half an hour and then at three, five, and eight, hours. This can best be done by having a second cloth lined tray placed on top of the hoops, then firmly holding both trays, turn over.

14. Leave overnight.

15. Next morning, take the cheese from the hoops and place into a cold 20% brine solution for 20 minutes to one hour, depending on the size of the cheese. A cheese of 125 grams needs about 20 minutes, and a 250 gram cheese needs about one hour.

$$pH \sim 5.0 - 5.1$$

(To make the brine add 200 gm salt to 800 mL of boiled water and allow to cool. Place brine in the refrigerator to chill the brine before use.

16. Remove from the brine.

17. Place on a rack to dry for 24 hours at room temperature.

18. Put the cheese into a humid environment at 11 to 15°C. The cheeses need to be turned and washed as often as necessary to prevent them drying out. This may be every second day or even daily.

The solution used to wash the cheese is a 4% brine solution, containing a special bacteria known as Brevibacterium linens. After three to four weeks, the rind develops a sticky surface growth(see below) that may vary from orange to reddish in colour.

During this period the cheese should be stored on a wooden shelf and turned after each wash. This method results in a cheese different in flavour than the previous method. The cheese will have an aroma that you may consider to be less than pleasant, but this is normal, and is not reflected in the flavour of the cheese.

19.　Allow the cheese surfaces to dry then wrap in plastic wrap and consume within 2 weeks.

Note: Some white mould may also be added to the milk or sprayed onto the cheese after brining.

WASHED RIND – SEMI HARD
****

1.   Pasteurise the milk and bring it to a setting temperature of 32°C. Add 150 mL of prepared Type B starter for each 10 litres, i.e. 1.5%. Mix in well.

2.   Add rennet at a rate of two and a half mL for each 10 litres of milk. Dilute the rennet with at least 10 times its volume of cool boiled water, i.e. 25 mL of cool boiled water to each two and a half mL of rennet. Then pour the diluted rennet immediately into the milk, taking care to pour it over as much of the surface as possible, stirring all the time while pouring it in. Mix in well for no less than one minute and no more than three minutes. Maintain the setting temperature.

3.   Allow the milk to set. This should take 30 to 40 minutes.

4.   Cut the curd into 13 mm cubes, then let stand for five minutes before stirring.

5.   Stir the curd gently and regularly over the next 40 to 50 minutes, being sure to maintain the temperature at about 32°C.

6.   Allow the curd to settle and remove whey to half the original milk level.

7.   Cooking: Slowly add two litres of boiled water cooled to 60°C. Add the water over 15 minutes, and at the end of this time the temperature of the curd/whey mixture should have just reached 38°C. If the temperature does not reach 38°C, you will need to heat by one of the indirect methods (see section on cooking the curds and whey). Stir frequently during this time.

8.   Stir the curd frequently over the next 30 mins. During the cooking process the curd becomes firmer, thus the intensity of the stirring can be increased without shattering the curd.

9.   Take off the whey to the point where it just covers the surface of the curd. Collect some of this whey as you may need to use it to cover the curd in the next step.

10.   Place the hoop that you are going to use to press and shape the cheese, into a slightly larger vessel. Put the curds and whey/water mix into the hoop, and press the curd down slightly while it is under the whey. At this stage, the whey must be covering the curd completely. If it doesn't cover the curd, you can add some of the spare whey that you collected. Press the curd for 15 minutes. A large two litre jar filled with water will have enough weight for pressing at this stage.

11.   Drain the whey from the outside vessel, while being sure not to disturb the curd in the hoop. Press the curd overnight. A weight of 10 kilograms should be sufficient.

12.   Next morning, place the cheese in a cold 20% brine solution (i.e.. 200 gm salt to each 800 mL of cool boiled water). The amount of time the cheese requires in the brine depends on the size of the cheese. As a guide, 250 gram requires about two hours, 500 gram requires about four hours, and 1 kilogram requires about eight hours.

13.   Remove the cheese from the brine, and allow to dry. It will take a minimum of one day to dry, perhaps longer.

14.   Put the cheese into a humid environment at 11 to 15°C. The cheeses need to be turned and washed as often as necessary to prevent them drying out. This

may be every second day or even daily. The solution used to wash the cheese is a 4% brine solution, containing a special bacteria known as Brevibacterium linens. After three to four weeks, the rind develops a sticky surface growth that may vary from orange to reddish in colour. During this period the cheese should be stored on a wooden shelf and turned after each wash. This method results in a cheese different in flavour than the previous method. The cheese will have an aroma that you may consider to be less than pleasant, but this is normal, and is not reflected in the flavour of the cheese. Cease the washing when the cheese is covered with the smear.

15. Allow the cheese to dry by removing from the humid area. Keep the cheese cool during this time.

16. When dry, apply two coats of plastic cheese coating, followed by a coat of wax.

17. Store at 10 to 15°C for two to six months, turning the cheese two to three times per week.

WHITE WENSLEYDALE
***

This cheese is similar to Cheshire and ranges in size from one and a half to five kg. It is a cheese that was first made in the Yorkshire dales by monks that came to England with William the Conqueror. The reason for it being called white Wensleydale, is that traditionally, Wensleydale is a blue veined cheese.

Sheep's milk was used initially, but later cow's milk gained favour. Traditionally raw milk was used without starter, and the milk was coagulated by a piece of calf's stomach. It is reported that when the calf stomach was unavailable, a black snail was used in the milk instead.

1. Pasteurise milk and bring it to a setting temperature of 30°C. Add 50 mL of prepared Type A starter for each 10 litres, i.e. 0.5%. Mix in well. Allow the milk to sit for a ripening period of 45 minutes, maintain the temperature.

2. Add rennet at a rate of two and a half mL for each 10 litres of milk, diluting it with at least 10 times its volume of cool boiled water, i.e. 25 mL of cool boiled water to each two and a half mL of rennet. Next pour the diluted rennet immediately into the milk, taking care to pour it over as much of the surface as possible, stirring all the time while pouring it in. Mix in well for no less than one minute and no more than three minutes. Maintain the setting temperature.

3. Allow the milk to set. This should take 40 to 45 minutes.

4. Cut the curd into 13 mm cubes, then allow the curds and whey to stand for five minutes.

5. Gently stir the curds and whey for 10 minutes, then let it sit for 15 to 20 minutes. Stir the curd again for the next 20 minutes as you raise the temperature to 32°C.

6. Stir the curd as often as required to prevent the curd from knitting together. This should continue until about two hours after the cutting step.

7. Drain off the whey and ladle the curd into a cheesecloth, tie in a bundle, opening the cloth every 10 to 15 minutes to cut or break the curd. After two hours the curd can then be cut into finger size pieces and salted. About 20 grams of salt per kilogram cheese is needed.

8. Place all the curd into a cheesecloth lined hoop and press overnight. Trim the excess cloth from the cheese.

9. The cheese should be stored in a cool damp area for three weeks and turned daily. If the cheese becomes too dry, you may need to apply a plastic cheese coating and wax. The cheese may be eaten at three weeks or aged for up to three months.

YOGHURT CHEESE
*

This is a very simple cheese, in fact, it is hardly even worth the one star rating. It is simply a concentrated form of yoghurt, quite sharp in flavour. It is an ideal base for dips. This technique can also be used to make Greek style Yoghurt.

1.   Either make yoghurt as per the yoghurt recipe, or buy some natural yoghurt.

2.   Very gently transfer the yoghurt to a cheesecloth lined colander. Tie the corners of the cloth together to form a bag and hang over a sink to drain at least 12 hours. During this time whey will drain out of the yoghurt and it will become drier. It is up to you to decide how long to drain. If you want a soft yoghurt cheese for dips, then less than 12 hours will be quite okay. If you want a firmer cheese, then drain for longer. These times will be affected by the temperature during draining. Warmer temperatures speed up drainage, and vice versa. As an alternative to hanging the bag to drain, a plate may be placed on top of the bag, and then a weight placed onto the plate. The whey will then be squeezed out by the pressure from the weight.

3.   Salt or herbs and spices can be added to enhance the flavour if desired.

A slight variation on this cheese is to make some Acidophilus/Bifido cultured milk and use instead of yoghurt.

144

RECIPES - OTHER PRODUCTS

CULTURED BUTTERMILK
**

There are two ways you can make cultured buttermilk. Method 1 is the simple and typical commercial method in Australia whereas method 2 is the traditional European method.

Method 1
1.   Standardise the milk by mixing whole milk and skim milk to achieve a fat level of about 1%. This can be done by mixing three parts of skim milk with one part of whole milk.

2.   Pasteurise the milk by heating it to 85°C, holding at that temperature for 10 minutes, then cooling it to 20 to 25°C.

3.   Inoculate with either one eighth of a teaspoon of Type B DVS starter powder, or 20 mL of prepared starter per litre of milk.

4.   Maintain the milk at 20 to 25°C until it has set, cool in the refrigerator overnight then stir well before consuming.   pH ~ 4.3 – 4.5

Method 2

1.   Pasteurise whole milk by heating to 80°C.   Cool immediately to 20 to 25°C.

2.   Inoculate with half a teaspoon of Type B DVS starter powder, or 20 mL of prepared starter per 10 litres of milk.   Maintain the temperature at 20 to 25°C until the milk sets. This should take between 16 and 24 hours.

3.   Churn the milk in a butter churn until the butter forms. Stop the churn and drain the buttermilk off. Fill into containers and refrigerate. Smaller quantities can be churned by whipping the milk in a food processor until the butter and buttermilk separate. The butter can be collected worked as described in the cultured butter recipe, but you will only get about half a kilogram from each ten litres of milk used.

SOUR CREAM
*

This sour cream recipe is very simple and produces a delicious product. You can use the normal 35% fat cream from the supermarket preferably long life or UHT cream. If you use cream containing preservatives, your efforts will be in vain. If you use long life cream, no heat treatment will be required, as it has already been heat treated. In this case start at step two.

To make a light sour cream, mix equal quantities of cream and wholemilk together. Any milk used to dilute the cream to make reduced fat sour cream, must be added to the cream before the heating commences in step one. If you use long life cream blended with long life milk, no heat treatment will be required, as it has already been heat treated. In this case start at step two.

1.   Before making sour cream, any raw cream must be heated to 90°C and then cooled to 25°C.

2.   Take 250 mL cream, and bring to 25°C. Add approximately one eighth of a teaspoon of Type B DVS starter powder or 20 mL of prepared starter. Mix in well.

3.   Keep at room temperature until cream sets, which may be one to two days. After this time you can continue to keep the cream warm until the flavour develops to your satisfaction. This is done by occasionally sampling to determine the flavour, then storing in the refrigerator when the flavour suits your liking.

TIKMILK
*

Tikmilk is our own name for a delightful cultured milk drink that we have developed. It is not as sour as yoghurt and is very simple to make. The secret to Tikmilk is the starter. It is a specially selected strain that is acid sensitive. As a result it does not produce a really sour milk, more a mild cultured flavour. This is particularly suitable to those people that consider natural yoghurt too sour.

1. Warm one litre of UHT milk up to 37°C and place into a clean sanitised container.

2. Place one eighth of a teaspoon of DVS Type E starter powder into the milk. Mix well until dissolved.

3. Maintain at 33 to 37°C until the milk has set.
   pH ~ 4.6

4. Place in a refrigerator to cool. When it is cold, shake vigorously to aerate and break up the contents. It is now ready to drink.

A delicious variation is to add freshly pureed fruits to the Tikmilk just prior to consumption. When making a prepared Type E starter for the modern version of Camembert the left over starter is Tikmilk.

Once you have made the Tikmilk, you can take a teaspoon of it out and use as your starter for your next batch. See starter notes for more details.

YOGHURT
*

Yoghurt is simple to make if you follow some basic rules. When making yoghurt, the milk should always be given a substantial heat treatment. The purpose of this is to destroy the spoilage bacteria present in the milk. With these bacteria gone, the yoghurt starter has no competition and can produce the lactic acid and flavour that is typical of yoghurt. The heat treatment also changes the proteins a little so that the body of the yoghurt is more viscous (thicker). Without heating, the yoghurt will be weak bodied and will not have a typical flavour.

The addition of milk powder to the milk will provide a thicker yoghurt and can be added at your discretion to suit your preferences. The heat treatment should always be done after the addition of powder, so that any bacteria in the powder are also destroyed.

If you are heating in a saucepan, the heating must be done carefully so that it does not burn on the base of the pan. If this happens then your yoghurt will have a cooked or burnt flavour.

Starter addition must not be done until the temperature is under 45°C. If starter is added with the milk too hot then the starter may be killed. It is wise to mix the starter in well, so that it is evenly distributed throughout the milk.

When incubating the yoghurt, the temperature is very important. The two bacteria in the yoghurt starter have different characteristics, which influence the yoghurt flavour and body. If the temperature of the milk is too warm during incubation, then one of the bacteria in the starter will proliferate and the other will not. This will result in an imbalance in the starters, producing a weak bodied yoghurt with a strong acid flavour. The best results are found by incubating the yoghurt at precisely 42 to 43°C. The incubation time should be about four to six hours. European yoghurts are often incubated at much lower temperatures (32 to 35°C), but the time of incubation will to be closer to 12 to 16 hours.

The yoghurt will coagulate during the incubation, but it should not be placed in the refrigerator until at least one hour after coagulation. The time allowed after the yoghurt coagulates and before refrigeration will determine its flavour intensity. To make a stronger flavoured yoghurt, simply leave it longer before cooling. To make a milder yoghurt, refrigerate soon after coagulation.

1. To each litre of whole milk, mix in 30 gm of skim milk powder. Alternatively place UHT milk into a sanitised jar and proceed to step four.

2. Heat to 90°C. A microwave oven is useful to heat small quantities of milk. If heating in a saucepan on the stove, stir continually to avoid burning milk on the base of the pan.

3. Upon reaching 90°C, remove the milk from the heat source, cover and allow to cool to 42 to 43°C. You may chose to remove the skin that will form, or stir it back into the milk.

4. Add approximately one tenth of a teaspoon of Type C DVS yoghurt starter powder or a teaspoon of your last batch of yoghurt.

5. Maintain the temperature at 42 to 43°C.

6. One hour after the yoghurt sets, (pH ~ 4.5) place in the refrigerator, and leave overnight. If a stronger flavour is preferred, leave longer before refrigerating. If a milder flavour is required, refrigerate as soon as it has set. Final pH ~ 4.2 – 4.4

7. Add fruit and sugar to taste. Do not use fresh fruit unless eating immediately. If you add fresh fruit and store the yoghurt, enzymes in the fruit will break down the protein in the yoghurt. The resultant yoghurt will most likely have a runny watery consistency.

•   To make a non-fat yoghurt, use one litre of skim milk and add 60 grams of skim milk powder.

•   To make a reduced fat yoghurt, use half a litre of skim milk, half a litre of whole milk and add 60 grams of skim milk powder.

•   If you are making yoghurt frequently, you may choose to use some of your freshly made yoghurt as your starter for your next batch, and so on for about three batches. This is suitable if the yoghurt is less than one week old. After using this technique a number of times, the two bacteria in the starter may get out of balance with each other, and your yoghurt may start to change flavour and body. See starter notes for details on subculturing.

GREEK STYLE YOGHURT
*

Greek style yoghurt can be made by one of two methods. The first is to make the yoghurt as per the yoghurt recipe and then drain the yoghurt through a cheesecloth until the yoghurt is thick enough for your liking. Usually it is rained to half its original volume. The second method requires 100 grams of skim milk powder to be added to each litre of milk and make the yoghurt as normal. It may take a couple of hours extra to set.

ACIDOPHILUS/BIFIDO CULTURED MILK
ACIDOPHILUS MILK
**

The production of cultured milk using starters Lactobacillus acidophilus and Bifidobacterium species is mainly confined to Scandinavian countries. These two bacteria have been isolated from the human digestive tract. They are claimed to have positive health effects on the human body. Whereas the two bacteria for making yoghurt do not survive passage through the stomach and intestinal system, these bacteria do. The benefits claimed by consuming this type of cultured milk include reduced flatulence, improved regularity and a positive effect on controlling harmful bacteria in the digestive system.

This recipe is the same for both Acidophilus/Bifido cultured milk and Acidophilus milk, except at step three. In step three, use Type F starter for Acidophilus/Bifido cultured milk and starter Type D when making Acidophilus milk.

It is essential to maintain strict hygiene and cleanliness when making this product. The milk takes about 24 hours or more to set because the bacteria grow very slowly. If the

152

milk becomes contaminated by other bacteria, the contaminants will probably grow faster than the starter spoiling the product. To ensure sterile milk we suggest that you use long life (UHT) milk. If you decide to use any other milk, then proceed from step one, otherwise place the UHT milk into a sanitised jar and proceed to step three. The jar must be sanitised thoroughly with sodium hypochlorite (see sanitisers section) or heat.

1. To each litre of whole milk, mix in 30 gm of skim milk powder. Alternatively place warm (38°C) UHT milk into a sanitised jar and proceed to step four.

2. Heat the milk until it boils. A microwave oven is useful to heat the milk. If heating in a saucepan on the stove, stir constantly to avoid burning the bottom of the pan.

3. Allow the milk to cool slowly to 38°C.

4. Add approximately one eighth of a teaspoon of Type D or F Starter powder(see page 106).

5. Maintain the temperature at 38°C until the milk sets. This will take approximately 24 hours. pH ~ 4.5

6. When the milk has set, place it in the fridge to cool.

7. When it is cool, consume as you would for yoghurt or stir well to liquefy if you prefer it as a drink.

• Because the milk takes so long to set, cleanliness and correct sanitation are essential!!!

ABT YOGHURT
**

The cultures used here are Lactobacillus **a**cidophilus, **B**ifidobacterium species and Streptococcus **t**hermophilus. The yogurt gets its aBt tag from the names of the starters, (note the bold letters in each bacterial name).

This yoghurt is a safer process in that the fermentation takes around 8 hours instead of the 24 hours for aB yoghurt. The shorter time means less chance of contaminant bacteria growing in competition with the aB cultures. The down side is that the numbers of acidophilus and Bifido bacteria are diminished by the growth of the t (streptococcus thermophilus) culture. If you want a good dose of aB cultures then this yoghurt when eaten fresh is a good source. It is essential to maintain strict hygiene and cleanliness when making this product.

1.   To each litre of whole milk, mix in 30 gm of skim milk powder. Alternatively place warm (38°C) UHT milk into a sanitised jar and proceed to step four.

2.   Heat the milk until it boils. A microwave oven is useful to heat the milk. If heating in a saucepan on the stove, stir constantly to avoid burning the bottom of the pan.

3.   Allow the milk to cool slowly to 38°C.

4.   Add approximately one eighth of a teaspoon of Type C aBt culture.

5.   Maintain the temperature at 38°C until the milk sets. This will take approximately 8-10 hours. pH ~ 4.5. The temperature must be carefully controlled to obtain maximum numbers of the beneficial bacteria.

6.   When the milk has set, place it in the fridge to cool.

7.   When it is cool, consume as you would for yoghurt or stir well to liquefy if you prefer it as a drink.

ABC YOGHURT
**

The cultures used here are Lactobacillus acidophilus, Bifidobacterium species and Lactobacillus casei. The yogurt gets its aBc tag from the names of the starters, (note the bold letters in each bacterial name). These are typically slow growing bacteria and therefore to minimise the chance of a contaminant bacteria growing either a Streptococcus thermophilus or the Type C yoghurt culture is added as well.

This yoghurt is a safer process in that the fermentation takes around 6-8 hours.

1. To each litre of whole milk, mix in 30 gm of skim milk powder. Alternatively place warm (38°C) UHT milk into a sanitised jar and proceed to step four.

2. Heat the milk until it boils. A microwave oven is useful to heat the milk. If heating in a saucepan on the stove, stir constantly to avoid burning the bottom of the pan.

3. Allow the milk to cool slowly to 38°C.

4. Add approximately one eighth of a teaspoon of Type C aBt culture and an equal quantity of Lactobacillus casei culture.

5. Maintain the temperature at 38°C until the milk sets. This will take approximately 8-10 hours. pH ~ 4.5. The temperature must be carefully controlled to obtain maximum numbers of the beneficial bacteria.

6. When the milk has set, place it in the fridge to cool.

7. When it is cool, consume as you would for yoghurt or stir well to liquefy by shaking if you prefer it as a drink.

CULTURED BUTTER
**

Cultured butter is very popular in European countries, and is usually consumed unsalted or lightly salted. In Australia, sweet cream butter is preferred, and only small quantities of cultured butter are consumed. It is however, slowly increasing in popularity.

1. Pasteurise one litre of fresh cream by heating to 80°C, then cool it to 18°C.

2. Add one quarter of a teaspoon of Type B starter (Direct Vat Set) or 50 mL of prepared starter and mix in well.

3. Keep the cream at room temperature for about 12 to 16 hours.   pH ~ 5.5

4. Cool in refrigerator overnight, then churn into butter. If you do not have a butter churn then the cream can be churned in a food processor with a whipping attachment. The cream goes through the whipped cream stage until there is free buttermilk and all the cream has been converted into butter grains. Work the butter grains with either butter pats or clean hands to remove the buttermilk. Finally shape the butter and refrigerate. Normally cultured butter is not salted, but a small amount of salt may be worked in before shaping.

• An alternative shorter method is to add 15% ready prepared Type B starter to cold pasteurised cream, (150 mL per litre) and churn immediately.

A traditional butter churn

PROBLEM SOLVING

One thing that is certain in cheesemaking - there will be a time when things don't work out quite right. We have never known a cheesemaker who hasn't had his/her share of problems. You will find that as you gain in experience you will have less problems. Not all your mistakes will be a disaster, and you might discover a new version of the cheese through some of them. It is important to use the temperatures and times as stated in the recipe and most importantly keep a record of what happened during the cheesemaking process. In most recipes you will be using starters which will produce acid. These starters are living organisms and may vary in activity depending on the following: quantity of starter, temperatures used, milk quality, antibiotics in the milk, age of the milk, hygiene level during cheesemaking and levels of other bacteria in the milk. These are just some of the reasons for variability in cheese.

Problem	Possible cause and remedy
Cheese milk won't coagulate after rennetting	Milk is too cold for the rennet. *Adjust milk to correct temperature.*
	Not enough rennet added to the milk.
	Add correct quantity.
	Not enough time allowed for the rennet to act properly. *Ensure the temperature and time specified in the recipe is used*
Milk won't coagulate	Rennet is too old and weak.

158

after rennetting(cont.)

If it takes too long to set the milk, or the curd is soft and weak, replace rennet with fresh stock.

The rennet is being diluted with chlorinated water instead of cool boiled water.
Presence of chlorine in the water destroys the rennet. Use only chlorine free water.

The rennet was diluted too early.
Ensure dilution of the rennet takes place only a few minutes before addition to the milk.

The milk is too hot when the rennet is added.
Ensure that milk is cooled to correct setting temperature before rennetting.

Rennet has been diluted with hot water.
Water must be cool. Cool boiled water is preferred (below 40°C).

Rennet has not been stirred into milk effectively.
Ensure proper stirring to mix rennet in well.

Rennet has not been diluted with water and does not mix into the milk effectively.
Dilute rennet with at least 10 times its volume with cool boiled or distilled water. Always dilute

the rennet just prior to adding it to the milk.

Milk has become diluted with too much added water.
Unlikely to occur if reasonable care is taken.

UHT or long life milk was used.

Long life milk will not coagulate with rennet.

Milk has been pasteurised at too high a temperature or for too long a time.

If you have used correct pasteurising time, check the thermometer for accuracy.

Starter won't set after incubation

The starter has been incubated at a temperature either too high or too low.
If the temperature has been too high then you will need to start again. If the temperature is too low then warming the milk should speed up the setting time.

Starter powder used is too old and no longer contains living bacteria.
Obtain fresh starter stocks.

The milk used to make the liquid starter possibly contains levels of

antibiotics.
Avoid using milk from animals under treatment with antibiotics. Refer to withholding times specified in the instructions with the antibiotics.

Insufficient time allowed for setting.
Starter types A and B both should be grown over 16 to 24 hours when incubated between 20 and 25°C. Other starters C and E require approximately 37°C and take 8 to 16 hours to set. Starter types D and F should not be used as starters. Add these directly to the milk as specified in the recipe.

Insufficient DVS starter powder or liquid starter has been used for the given quantity of milk.
Add more starter next time.

The starter was placed into the starter milk whilst the milk was still hot.
Allow milk to cool to the specified optimum temperature range for the starter type before adding starter. Consult optimum temperatures for each starter.

Cheese has a bitter flavour	Too much rennet is added to the milk. *Measure rennet very carefully adding exact quantities. A small syringe is ideal. If the rennet is double or quadruple strength, then only half or a quarter of the specified amount is needed.*
	Not enough salt has been added to the cheese at the salting step. *Increase salting rate next time.*
	Milk used for cheesemaking is poor quality, i.e. contaminated by undesirable bacteria. *Only use milk less than three days old for making cheese. Ensure that all equipment is clean and sanitised before use.*
Cheese has a rancid flavour	Milk used for cheesemaking is poor quality, i.e. contaminated by undesirable bacteria. *Only use milk less than three days old for making cheese. Ensure that all equipment is clean and sanitised before use.*
	Raw milk has been used for making cheese. *Pasteurisation will prevent this problem.*
Cheese has a putrid flavour	Starter was not used or it did not work. *Add starter. Ensure that if using liquid prepared starter it is used*

within 24 hours of being made.

The milk used for cheesemaking contains antibiotics.
Avoid using milk from animals under treatment with antibiotics. Refer to withholding times specified in the instructions.

Milk used for cheesemaking is poor quality, i.e. contaminated by undesirable bacteria.
Only use milk less than three days old for making cheese. Ensure that all equipment is clean and sanitised before use.

Cheese is too acidic

Too much starter is added to the cheese milk.
Use less starter next time.

Cheese contains too much moisture from low cooking temperatures, or not enough stirring during the curds and whey stage.
Ensure correct cooking rate and temperature is used. If these are okay then increase the stirring of the curds and whey.

Too little salt has been added to the cheese.
Increase salting rate next time.

The curds have been left in the whey for too long.
Stick closely to specified times.

163

Cheese has a weedy or feedy flavour	From the animals feeding on strong smelling pastures or weeds e.g. clover, onion weed, capeweed or lucerne. *Remove animals from source of feed at least four hours before milking.*
Cheese has a fruity flavour	Stray bacteria or yeasts have contaminated the milk. *Ensure that all equipment is clean and sanitised before use.*
	Poor hygiene used during milking or cheesemaking. *Ensure that hands and arms are clean and sanitised before milking and cheesemaking.*
	Dirty milking or cheesemaking equipment used. *Ensure that all equipment is clean and sanitised before use.*
Cheese has a chemical flavour	Too much residual sanitiser in the milk. *Drain all sanitiser from utensils before contact with the milk.*
	Cheese absorbs odours from other foods or chemicals stored in the same area. *Avoid storing cheese in presence of strong odours.*

Cheese too soft/moist	Coagulum not cut into small enough pieces. *Next time try a finer cut.*
	Curds and whey are cooked at too low a temperature. *Ensure that the correct cooking temperatures are used and the thermometer is accurate.*
	Not enough stirring during the curds and whey stage, i.e. the time between cutting the curd and whey removal *Increase stirring rate or time.*
	The fat level of the milk is too high. *Standardise the milk to remove some fat.*
	Insufficient pressing pressure or time. *Use heavier weights and/or press for longer.*
Cheese too hard or dry	Curd is cut into pieces that are too small. *Next time try a slightly wider cut.*
	Cooking temperatures used are too high. *Reduce the cooking temperature and check accuracy of the thermometer.*
	Curds and whey were over stirred or stirred for too long.

Reduce the stirring intensity or cut back the stirring time.

Cheese has not been waxed or wax coating is too thin.
Apply a wax coating and ensure that the wax is not too hot. If the coat is too thin, then allow the wax to cool and re-coat.

Storage area does not have enough humidity.
Place tray of water in storage area. Drape cloth into water to act as a wick, and/or spray atomised water into the storage area.

Cheese is too small in size or too flat.
Use more milk to make a bigger cheese or change shape by using different hoops.

Cheese contains tiny holes

The cheese has been infected by coliform bacteria from poor hygiene or cleaning.
Ensure that all equipment is clean and sanitised before use. Ensure that hands and arms are clean and sanitised before milking and cheesemaking.

Curd won't knit together

Curd is too cold at the hooping stage.
Prevent the curd from cooling off during the time from removing the whey until hooping.

	Not enough pressure has been applied during pressing. *Press longer or use more weight.*
	Curd is too acidic or too dry. *See causes and remedies for acid and/or dry cheese.*
Unwanted mould on the surface of the cheese	Humidity is too high during storage *Unable to correct if the problem is humidity from the atmosphere.*
	Unclean conditions in the storage area or making room. *You will need to thoroughly clean and sanitise your work and cheese store rooms. Avoid exposing the cheese in the refrigerator.*
Unwanted moulds on Camembert and Brie	Dirty conditions in the maturing area or making room. *You will need to thoroughly clean and sanitise your work and cheese store rooms.*
White mould grows too slowly	Low humidity during storage. *Place tray of water in storage area. Drape cloth into water to act as a wick, and/or spray atomised water into the storage area.*
	Temperatures too low during storage. *Increase the temperature to that specified in the recipe.*
	Not enough mould spores added

	to the milk, or brine.
	Add more spores next time.
Cheese surface is greasy	Overheating of the milk or curds and whey.
	Use correct temperatures and check thermometer accuracy.
	Over agitation of the curds and whey.
	Reduce stirring intensity.
	Cheese store room is too warm.
	Lower the storage temperatures.
Wax won't adhere to the cheese	Cheese is too damp at time of waxing.
	Allow more time for the cheese to dry.
	Plastic cheese coating was not used before waxing.
	Apply one or two coatings of plastic cheese coating then wax.
	Cheese has greasy surface *(see previous defect).*

GLOSSARY

Acid Curd	The gel-like state that milk is brought to, when a high level of acidity is achieved. The acidity is produced by the activity of starter bacteria, and it precipitates the milk protein into a solid curd.
Acidity	The amount of acidity (sourness) in the milk. Acidity is an important element in cheesemaking and it is produced by cheese starter bacteria.
Ammoniated	A mould or washed rind cheese that is overripe and smells or tastes strongly of ammonia.
Annatto	A natural vegetable extract (from the seeds of a South American bush Bixa orellana) which is used to colour cheese yellow, orange or red.
Bacteria	Microscopic single cell organisms found almost everywhere. Lactic acid-producing bacteria are useful and essential in the production of most cheeses.

Bacteria linens	A red bacteria which is encouraged to grow on the surfaces of cheeses like Brick, Havarti, Tilsit and Limburger, to produce a characteristic flavour. The full name of the organism is Brevibacterium linens.
Bacterial surface ripened cheese	Cheese that has a heavy growth of bacteria on the surface, to produce a distinct flavour. Brick and Limburger are examples of bacterial surface ripened cheeses.
Blind	Usually refers to a Swiss cheese types that have no holes or eyes in it.
Body	The inside of a cheese which is assessed by graders using terms such as, firm, weak, pasty, flaky, close, short etc.
Brine	A mixture of salt and water. Cheese salt, rock salt or table salt can be used. Do not use iodised salt.
Brined cheese	A cheese immersed in a brine solution.

Butterfat

The fat portion in milk. Butterfat can vary from 3.6 to 6.6 percent of the milk. Sometimes called milk fat.

Calf Rennet

Calf rennet is derived from the fourth stomach of a milk-fed calf. It contains the enzyme Chymosin which has the ability to coagulate milk. Animal rennet is commonly available in liquid form.

Cheddaring

The process during Cheddar cheesemaking after the whey is drained from the curds. The curds are then kept warm for approximately 90 to 120 minutes.

Cheesecloth

A coarse to fine cloth either cotton or plastic used to drain curds, line cheese hoops and other miscellaneous uses.

Cheese Salt

A coarse flake salt. Salt not iodised is the most desirable type to use in cheesemaking.

Cheese
Starter Culture

A bacterial culture added to milk as the first step in making many cheeses. The bacteria produce acid in the milk and in the curds. There are different categories of starter culture; mesophilic, thermophilic, gas producing and aroma producing.

Cheese Wax

A pliable wax, usually a mixture of paraffin and microcrystalline wax, with a low melting point, which produces an airtight seal and a moisture barrier around the cheese. It is applied to rinded cheeses before maturing, or to other cheeses prior to sale to give an attractive appearance.

Coagulation

The solidification of milk through the action of acid and/or enzymes. The enzymatic method uses a product known as rennet.

Cooking

A step in cheesemaking during which the cut curd is heated to assist in whey removal from the curds.

Curd	The coagulated part of milk, consisting of solid protein with some fat, lactose and residual whey.
Cutting the Curd	A step in cheesemaking in which the curd is cut into equal sized pieces.
Draining	The step in cheesemaking in which the whey is separated from the curd.
Dry matter	All parts of a cheese excluding the moisture: i.e.. fat, protein, lactose and minerals.
Enzyme	The active component of rennet is an enzyme. An enzyme accelerates a reaction, in the case of rennet, the coagulating or setting of the milk.
Eyes	Small or large holes in a cheese produced after gas formation by selected bacteria.

Fat in Dry Matter	The proportion of fat in a cheese, expressed as a percentage of the dry weight (fat in dry matter), not total weight. Most cheeses have a fat in dry matter content of around 45-50%. The actual fat content of the cheese is usually in the range of 25-35%.
Grana	A general term for hard Italian grating cheeses like Parmesan.
Homogenisation	A mechanical process that reduces the size of the fat globules of milk, so that the cream will no longer rise to the surface of the milk.
Hooping	A step in cheesemaking during which the curd is placed in a cheese mould (hoop). The cheese hoop will help produce the final shape of the cheese, and assists in whey drainage from the cheese.
Lactic Acid	The acid produced in milk or curd during cheesemaking. Cheese starter culture bacteria break down the milk sugar(lactose), and produce lactic acid as a by-product.

Lactose	The sugar naturally present in milk. Lactose can constitute up to five percent of the total weight of milk. It is not a very sweet sugar compared to glucose.
Lipase	An enzyme added to milk to break down the fat and create piquant flavour
Maturing	A step in cheesemaking in which the cheese is stored at a particular temperature and/or relative humidity for a certain time, in order to develop its distinct flavour and/or for the body to breakdown.
Mesophilic Starter	Lactic acid-producing starter bacteria which is used to produce cheeses when the cooking temperature is 39°C or lower.
Microbial rennet	See vegetarian rennet.
Milling	A step in cheesemaking, during which the curd is diced into smaller potato chip like pieces before being salted.

Mould-ripened Cheese	A cheese which has matured with the assistance of a mould, growing on the surface or inside. Two types of mould are most common in cheesemaking. They are blue mould for the blue cheeses and white mould for Camembert and related cheeses.
Pasta Filata	Italian expression for plastic-curd cheeses, where thin strips of cheese curd are placed into a hot water bath and worked up until homogenous. Mozzarella is an example of a pasta filata cheese.
Pasteurisation	The heating of milk by either batch method i.e.. 61°C and holding it for 30 minutes or by a high temperature/short time method of 72°C and holding for 15 seconds. The aim is to destroy pathogenic organisms which may be harmful to humans. An alternative of heating the milk to 68°C and holding for one minute is usually more time efficient for the home cheesemaker.

Penicillium candidum	A white mould which is encouraged to grow on the surface of a number of soft mould-ripened cheeses including Camembert and Brie.
Penicillium roqueforti	A blue mould which is encouraged to grow on the surface and in the interior of a variety of blue cheeses, such as Stilton, Gorgonzola, Roquefort and Blue Vein.
Plastic curd	Generic term for cheese made by immersing the curd in hot water and working it until it becomes elastic and can be moulded into the shape required. (see also, pasta filata)
Pressing	A step in cheesemaking during which the curds are placed in a cheesecloth-lined hoop or mould, then placed under pressure to remove more whey and create a closer textured cheese.
Rennetting	The step in cheesemaking in which rennet is added to milk in order to bring about coagulation.

Ripening	A step in cheesemaking just after starter addition and before rennetting in which the milk is allowed to undergo an increase in acidity, due to the activity of cheese starter culture bacteria. Sometimes the term ripening is used to indicate maturing. It is less confusing to use the term maturing in this context.
Rind	Outer coating of a cheese formed by surface drying, often treated by rubbing, brining, oiling, blackening or other methods to produce the desired characteristics. Natural rinds are usually edible, synthetic rinds made by adding a layer of other substances may not be.
Salting	A step in cheesemaking in which coarse salt is added to the curds before moulding or to the surface of the finished cheese. Alternatively some cheeses are salted by immersion in a brine solution.

Soft Cheese	A cheese which is not pressed, contains a high moisture content, and is eaten very soon after production.
Stabilised Camembert	A term used to describe a style of Camembert where the pH is stabilised by the use of type E starter. Also referred to as Camembert (Modern) style.
Thermophilic Cheese Starter Culture	A type of bacterial starter which is used for the making of cheeses which have higher cooking temperatures, than that used when using mesophilic starters. Many Italian cheeses and Swiss cheese require a thermophilic culture.
Vegetarian Rennet	A product used to coagulate milk. It is produced by either yeast or bacteria. The enzyme is extracted from the cells then filtered and is free of any of the yeast or bacteria.

Whey	The liquid portion of milk which separates from the curds after coagulation of the milk protein. Whey contains water, milk sugar, whey proteins, minerals and some fat. It should be a clear greenish colour and not milky.
Whey Protein	Protein in milk which is not precipitated by the addition of rennet. Whey protein remains in the whey and can be precipitated by high temperatures to make Ricotta.

Log sheet for Cheesemaking Date: _____

Milk Details		Vat 1		
	Quantity-Litres			
	Fat %			
	Protein %			
	Casein %			
Starter Details	Supplier & Code			
	Quantity			
	Rennet Type Rennet Qty			
Manufacture details and comments		Time	Temp	Acid/pH
	Vat filled			
	Starter Added			
	Rennet Added			
	Curd cut			
	Knife size			
	Heat on			
	Heat off			
	1/2 whey off			
	Whey off			
	Dry Stirs (if any)			
	Milled			
	Salted			
	Hooped			
	Cloths pulled up			
	Salt %			
	Qty Salt			
Weights of Cheese				

SUPPLY OF CHEESEMAKING INGREDIENTS.

CheeseLinks

PO Box 146
Little River, Victoria, 3211
Australia,
Phone (03) 52831396
Fax   (03) 52831096
Website: www.cheeselinks.com.au
E-mail: cheesemaster@cheeselinks.com.au

Starters, rennet, moulds, aroma cultures, lipase, cheesecloth, thermometers, plastic cheese coating, cheesewax, cheese wraps, cheese baskets, cheese hoops, maturing racks, calcium solution, separators.

INDEX

Notes

Notes

Notes